Molecular Bioenergetics

ACS SYMPOSIUM SERIES **883**

Molecular Bioenergetics

Simulations of Electron, Proton, and Energy Transfer

Ralph A. Wheeler, Editor
The University of Oklahoma

Sponsored by the
ACS Division of Computers in Chemistry

American Chemical Society, Washington, DC

Library of Congress Cataloging-in-Publication Data

Molecular bioenergetics : simulations of electron, proton, and energy transfer / Ralph A. Wheeler, editor ; sponsored by the ACS Division of Computers in Chemistry.

    p. cm.—(ACS symposium series ; 883)

  Based on the symposium presented at the 218th American Chemical Society (ACS) national meeting in New Orleans, La., 1999.

  Includes bibliographical references and index.

  ISBN 0–8412–3720–4

  1. Bioenergetics—Congresses. 2. Electron transport—Congresses. 3. Proton transfer reactions—Congresses. 4. Photosynthesis—Congresses.

  I. Wheeler, Ralph A. II. American Chemical Society. Meeting (218th : 1999 : New Orleans, La.) III. Series.

QP517.B54M64   2004
572'.43—dc22   2004041121

The paper used in this publication meets the minimum requirements of American National Standard for Information Sciences—Permanence of Paper for Printed Library Materials, ANSI Z39.48–1984.

Copyright © 2004 American Chemical Society

Distributed by Oxford University Press

All Rights Reserved. Reprographic copying beyond that permitted by Sections 107 or 108 of the U.S. Copyright Act is allowed for internal use only, provided that a per-chapter fee of $27.25 plus $0.75 per page is paid to the Copyright Clearance Center, Inc., 222 Rosewood Drive, Danvers, MA 01923, USA. Republication or reproduction for sale of pages in this book is permitted only under license from ACS. Direct these and other permission requests to ACS Copyright Office, Publications Division, 1155 16th St., N.W., Washington, DC 20036.

The citation of trade names and/or names of manufacturers in this publication is not to be construed as an endorsement or as approval by ACS of the commercial products or services referenced herein; nor should the mere reference herein to any drawing, specification, chemical process, or other data be regarded as a license or as a conveyance of any right or permission to the holder, reader, or any other person or corporation, to manufacture, reproduce, use, or sell any patented invention or copyrighted work that may in any way be related thereto. Registered names, trademarks, etc., used in this publication, even without specific indication thereof, are not to be considered unprotected by law.

PRINTED IN THE UNITED STATES OF AMERICA

Foreword

The ACS Symposium Series was first published in 1974 to provide a mechanism for publishing symposia quickly in book form. The purpose of the series is to publish timely, comprehensive books developed from ACS sponsored symposia based on current scientific research. Occasionally, books are developed from symposia sponsored by other organizations when the topic is of keen interest to the chemistry audience.

Before agreeing to publish a book, the proposed table of contents is reviewed for appropriate and comprehensive coverage and for interest to the audience. Some papers may be excluded to better focus the book; others may be added to provide comprehensiveness. When appropriate, overview or introductory chapters are added. Drafts of chapters are peer-reviewed prior to final acceptance or rejection, and manuscripts are prepared in camera-ready format.

As a rule, only original research papers and original review papers are included in the volumes. Verbatim reproductions of previously published papers are not accepted.

ACS Books Department

Contents

Preface..ix

1. Introduction to the Molecular Bioenergetics of Electron,
   Proton, and Energy Transfer..1
   Ralph A. Wheeler

2. A Quantum Chemical View of the Initial Photochemical
   Event in Photosynthesis..7
   Michael C. Zerner and Ralph A. Wheeler

3. Molecular Modeling and Simulation of a Reaction Center
   Protein..37
   Matteo Ceccarelli, Marc Souaille, and Massimo Marchi

4. Simulating Thermochemistry of p-Benzo-quinone
   Reduction and Binding of Ubiquinone in the Photosynthetic
   Reaction Center...51
   Ralph A. Wheeler

5. Problems Evaluating Energetics of Electron Transfer
   from Q_A to Q_B: The Light-Exposed and Dark-Adapted
   Bacterial Reaction Center...71
   Björn Rabenstein and Ernst-Walter Knapp

6. Modeling the First Electron Transfer from Q_A to Q_B
   in Reaction Center Proteins from *Rb. sphaeroides*....................93
   E. G. Alexov and M. R. Gunner

7. Dynamics of Electron Transfer Pathways in Redox Proteins............107
   Ilya A. Balabin and José Nelson Onuchic

8. Ab Initio Calculations of Long-Distance Electron Tunneling
   in Proteins with the Method of Tunneling Currents..................119
   Jongseob Kim, Xuehe Zheng, Yuri Georgievskii,
   and Alexei A. Stuchebrukhov

9. **Proton-Coupled Electron Transfer Reactions: A Theoretical Approach** 145
   Robert I. Cukier

10. **Proton Relay in Membrane Proteins** 159
    Régis Pomès

11. **Computer Simulation of Energy-Transducing Proteins and Peptide:Membrane Interactions** 175
    Dan Mihailescu, G. Matthias Ullmann, and Jeremy C. Smith

Indexes

Author Index 189

Subject Index 191

Preface

Approximately 0.05% of all sunlight reaching the earth's surface is used by photosynthetic organisms to synthesize organic compounds. All other organisms use these compounds as energy sources for their metabolism and thus the conversion of light energy into biologically usable forms is critical to the viability of living organisms. Typically, energy is converted by organisms into cellular energy sources in the form of high-energy compounds such as ATP, ion gradients, or conformational energy of proteins or their cofactors. Bioenergetics encompasses the thermodynamics, kinetics, and molecular mechanisms of the processes important for biological energy storage and its use. Among the most elementary chemical reactions central to these processes are electron transfer, proton transfer, and proton-coupled electron transfer.

This volume grew out of an American Chemical Society (ACS) symposium titled *Bioenergetics*. The ACS Division of Computers in Chemistry sponsored the symposium, whose goal was to bring together scientists from different disciplines to discuss current achievements and future directions in molecular-level simulations of electron and proton transfer. This volume provides a sampling of recently developed simulation methods, as well as their applications to prototypical biochemical systems such as the photosynthetic reaction center and bacteriorhodopsin.

Contributors to the symposium and to this volume are thanked for sharing their work and expertise with their audiences. In addition, I am grateful to Stacy Van der Wall, Bob Hauserman, and Margaret Brown at the ACS Books Department for their help with this volume.

Ralph A. Wheeler
Department of Chemistry and Biochemistry
University of Oklahoma
620 Parrington Oval, Room 208
Norman, OK 73019

Molecular Bioenergetics

Chapter 1

Introduction to the Molecular Bioenergetics of Electron, Proton, and Energy Transfer

Ralph A. Wheeler

Department of Chemistry and Biochemistry, University of Oklahoma, 620 Parrinton Oval, Room 208, Norman, OK 73019 (telephone: 405-325-3502, e-mail: rawheeler@chemdept.chem.ou.edu)

Since all biological structures are thermodynamically less stable than simpler molecules such as carbon dioxide and water, life is a constant uphill battle against the forces of thermodynamics. In fact, the thermodynamics of life demands that all living organisms continually harvest energy from their environment, transform that energy into useful forms, and use stored energy in metabolic processes to maintain body temperature, move, grow, etc. Energetic challenges require organisms to store harvested energy for use at a time and place where they need it most. The structural and functional aspects of these processes of energy transformation in biological systems define the field of bioenergetics.[1,2]

On a molecular level, living organisms meet the demands of thermodynamics by capturing a continual supply of energy to build complex molecules, organize them, move them around, and, later, destroy them to release their stored energy. Primary environmental sources of energy typically include light or chemically reduced compounds that can later be oxidized to release stored energy on demand. Understanding the molecular structures and transformation involved in these processes of biological energy conversion defines "molecular bioenergetics".

Figure 1 provides a simplified overview of the major molecular-level processes involved in bioenergetics. In Figure 1, light energy is stored in the form of reduced cofactors or concentration gradients. The conversion is assisted by protein structures such as bacteriorhodopsin or generic "reaction centers". The energy-rich ATP molecule occupies a central role in energy storage and is synthesized from other, chemically reduced compounds. Finally, energy stored

© 2004 American Chemical Society

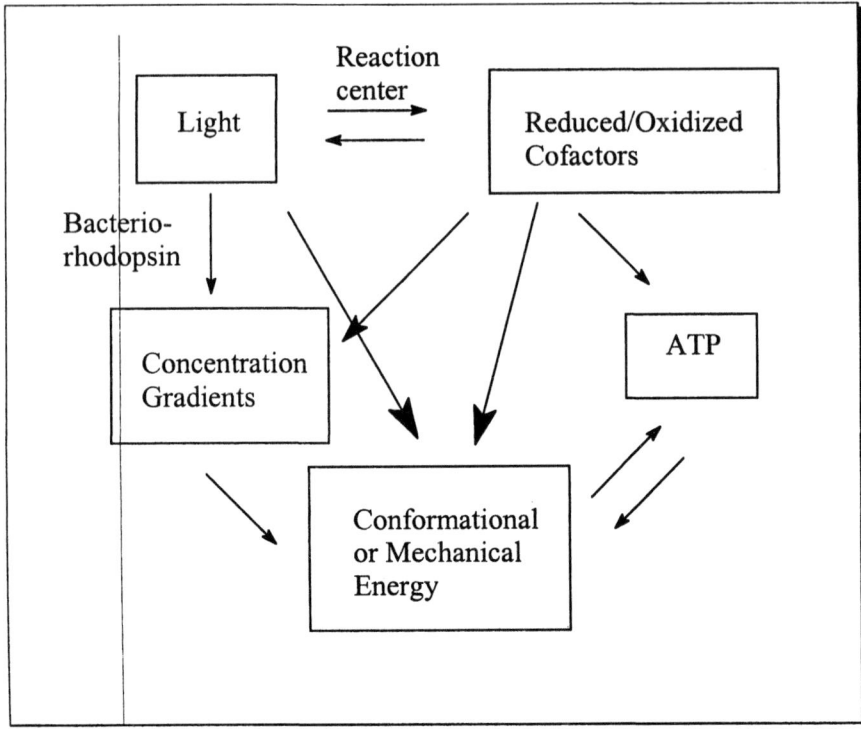

Figure 1. Simplified overview of major, molecular-level processes involved in bioenergetics. Reaction centers store light energy by reducing cofactors and bacteriorhodopsin converts light to a proton gradient. Eventually, concentration gradients and ATP provide biochemically useful energy.

in the form of concentration gradients or ATP may be released to perform useful functions.

The proteins involved in energy storage, transduction, and use are many and varied and perform such diverse functions as electron transfer, proton transfer, and covalent bond making and breaking. Figure 1 implies that the ultimate energy source for most organisms is solar energy stored by photosynthesis and converted to chemical energy in protein structures called "reaction centers".[3] Ultimately, the stored energy is released for an organism's use in organelles call mitochondria. In between, a host of proteins participate in energy transduction—the transformation of energy from one form or location to another. This volume provides a sampling of computational methods and their applications to model energy storage, transduction, and use. Since many of the proteins involved in bioenergetics are membrane-bound, and membrane proteins are notoriously difficult to extract from the membrane, crystallize, and structurally characterize, structural data that form the starting point for computational studies are difficult to collect. Therefore, simulations of bioenergetic processes focus on a relatively small number of proteins, including the bacterial photosynthetic reaction center (the first membrane protein characterized by X-ray crystallography)[4] and bacteriorhodopsin.[5]

The photosynthetic reaction center stores light energy by effecting electron transfer to reduce an electron transfer cofactor and form a proton gradient across the membrane. The arrangement of electron transfer cofactors is indicated in Figure 2 and includes a "special pair" of bacteriochlorophyll molecules, two accessory bacteriochloroophylls, two bacteriopheophytins, two quinone electron acceptors, and a non-heme iron. The reaction center functions

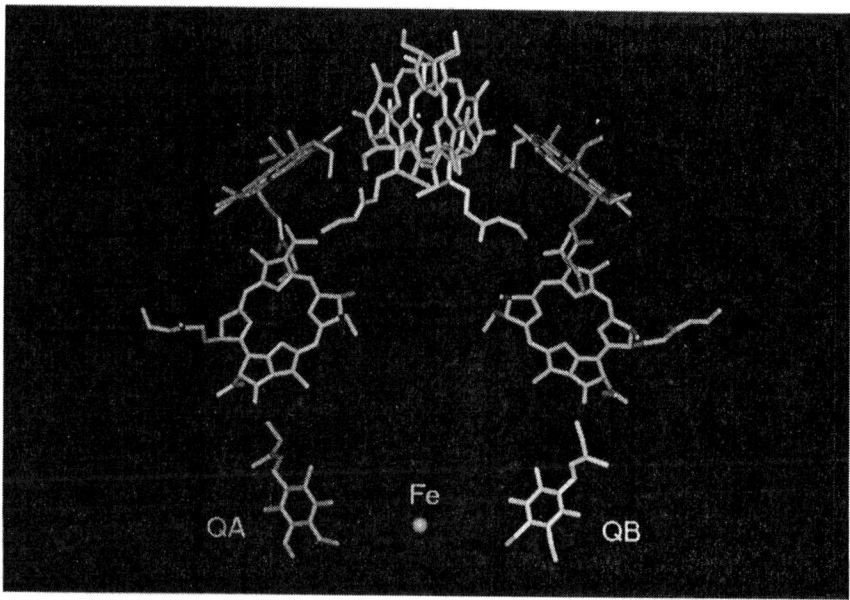

Figure 2. Arrangement of the electron transfer cofactors in the photosynthetic reaction center protein from the bacterium *Rhodobacter sphaeroides*. The figure shows the special pair of bacteriochlorophylls (top, in green and light blue), two accessory bacteriochlorophyll molecules (dark blue), two bacteriopheophytins (red), the primary quinone (Q_A), the secondary quinone (Q_B), and the non-heme iron.

(See page 1 in color insert.)

by accepting energy from antenna complexes that harvest sunlight. The energy electronically excites the special pair, which transfers a single electron to the primary quinone Q_A and then on to the secondary quinone Q_B. The reduced Q_B^- takes up one proton and the special pair is reduced by cytochrome c to reset the reaction center for a second electron transfer. A second electron is transferred from the special pair to further reduce Q_BH, which also accepts a second proton and then diffuses out of the reaction center to engage is subsequent redox chemistry. After reduction of the oxidized special pair by cytochrome c and replacement of the reduced Q_BH_2 with another quinone, the reaction center is reset for another two-electron reduction cycle. The overall reaction results in the two-electron reduction of the quinone Q_B and the creation of a proton gradient across the membrane. The proton gradient subsequently drives further energy storage reactions by making ATP from ADP.

Bacteriorhodopsin is the quintessential transmembrane ion pump.[5] It consists of a small, seven-helix protein where proton transport across the membrane is driven by photoisomerization of retinal from the all *trans* to the 13-*cis*,15-*anti* configuration. A number of high-resolution crystal structures of the protein and its photointermediates have been used to propose several competing mechanisms describing proton translocation to the extracellular surface. Unresolved issues include understanding how conformational changes couple to proton transfer[6,7] and the role played by water molecules in the proton transfer process.[8-10]

In this volume, Michael Zerner describes quantum chemical calculations to address design issues relevant to the photosynthetic reaction center, with special emphasis on the special pair. Ceccarelli et al. consider the electron transfer events immediately following electronic excitation of the special pair, whereas the next three contributions address the challenges of simulating quinone reduction in the photosynthetic reaction center. The next two contributions, by Balabin and Onuchic and by Kim, Stuchebrukhov et al. consider electron transfer pathways in proteins. Simulations of proton transfer pathways and rates are considered by Pomes, while Cukier presents methods of simulating proton-coupled electron transfer. Finally, Mihailescu, Ullmann and Smith summarize simulations of energy transduction in bacteriorhodopsin and emphasize the importance of peptide-membrane interactions.

Overall, this volume provides a sample of the challenges and opportunities inherent in computer simulations to gain a molecular-level understanding of bioenergetics. Challenges include the need for a starting structure, an appropriate physical model for the process of interest, computer software to implement the model, and a means of analysis to translate the simulation into an understanding of the process and to make useful predictions. While simulations to understand bioenergetics have been challenged by the scarcity of X-ray diffraction structures of the proteins involved, exciting new techniques to overexpress, extract, purify, and crystallize membrane proteins[11] have already given an abundance of new structures. In studies of photosynthesis alone, high-resolution X-ray structures of photosystem I,[12,13] photosystem II,[14] and the cytochrome b_6f complex[15] that provides a functional coupling between photosystems I and II, were recently published and promise an explosion of simulation work to understand the bioenergetics of oxygenic photosynthesis. Methods, models, and insights presented by contributors to this volume will undoubtedly lead the way in these and other, as yet unforeseen efforts to write a new "molecular bioenergetics" chapter in the history of bioenergetics.[16]

References

(1) Cramer, W. A.; Knaff, D. B. *Energy Transduction in Biological Membranes*; Springer-Verlag: New York, 1990.
(2) Nicholls, D. G.; Ferguson, S. J. *Bioenergetics 3*; Academic Press: Amsterdam, 2002.
(3) Blankenship, R. E. *Molecular Mechanisms of Photosynthesis*; Blackwell Science: Oxford, 2002.
(4) Fritzsch, G.; Kuglstatter, A. The structure of reaction centers from purple bacteria. *The Photochemistry of Carotenoids*; Kluwer: Dordrecht, 1999; pp 99-122 and references therein.
(5) Luecke, H.; Lanyi, J. K. Structural clues to the mechanism of ion pumping in bacteriorhodopsin. *Adv. Protein Chem.* **2003**, *63*, 111-130 and references therein.
(6) Balashov, S. P. Protonation reactions and their coupling in bacteriorhodopsin. *Biochim. Biophys. Acta* **2000**, *1460*, 75-94.
(7) Subramaniam, S.; Henderson, R. Crystallographic analysis of protein conformational changes in the bacteriorhodopsin photocycle. *Biochim. Biophys. Acta* **2000**, *1460*, 157-165.
(8) Cao, Y.; Varo, G.; Chang, M.; Ni, B.; Needleman, R. et al. Water is required for proton transfer from aspartate-96 to the bacteriorhodopsin Schiff base. *Biochemistry* **1991**, *30*, 10972-10979.
(9) Pebay-Peyroula, E.; Rummel, G.; Rosenbusch, J. P.; Landau, E. M. X-ray structure of bacteriorhodopsin at 2.5 angstroms from microcrystals grown in lipidic cubic phases. *Science* **1997**, *277*, 1676-1681.
(10) Belrhali, H.; Nollert, P.; Royant, A.; Menzel, C.; Rosenbusch, J. P. et al. Protein, lipid, and water organization in bacteriorhodopsin crystals: a molecular view of the purple membrane at 1.9 angstroms resolution. *Structure (London)* **1999**, *7*, 909-917.

(11) See, for example, Kyogoku, Y.; Fujiyoshi, Y.; Shimada, I.; Nakamura, H.; Tsukihara, T. et al. Structural genomics of membrane proteins. *Acc. Chem. Res.* **2003**, *36*, 199-206.

(12) Jordan, P.; Fromme, P.; Will, H. T.; Klukas, O.; Saenger, W. et al. Three-dimensional structure of cyanobacterial photosystem I at 2.5 angstrom resolution. *Nature* **2001**, *411*, 909-917.

(13) Saenger, W.; Jordan, P.; Krauss, N. The assembly of protein subunits and cofactors in photosystem I. *Curr. Opin. Struc. Biol.* **2002**, *12*, 244-254.

(14) Zouni, A.; Witt, H.-T.; Kern, J.; Fromme, P.; Krauss, N. et al. Crystal structure of photosystem II from Synechococcus elongatus at 3.8 angstrom resolution. *Nature* **2001**, *409*, 739-743.

(15) Kurisu, G.; Zhang, H.; Smith, J. L.; Cramer, W. A. Structure of the cytochrome b_6f complex of oxygenic photosynthesis: tuning the cavity. *Science* **2003**, *302*, 1009-1014.

(16) Robinson, J. D. *Moving Questions, A History of Membrane Transport and Bioenergetics*; Oxford University Press: New York, 1997.

Chapter 2

A Quantum Chemical View of the Initial Photochemical Event in Photosynthesis

Michael C. Zerner[1,†] and Ralph A. Wheeler[2]

[1]Quantum Theory Project, The University of Florida,
Gainesville, FL 32611
[2]Department of Chemistry and Biochemistry, University of Oklahoma, 620
Parrinton Oval, Room 208, Norman, OK 73019
†Deceased

1. INTRODUCTION

Photosynthesis is the process in which light energy is used to convert carbon dioxide, one of the most stable molecules naturally occurring, into sugars and starches. This process is responsible for essentially all of the carbon atomic fixation on our planet - for essentially all of the biomass on earth. Quite literally, the earth's atmosphere has been converted from one of reducing power to one of oxidizing power by photosynthesis. Without the efficiency of this process life as we know it could not exist.

The overall stochiometry of this process is given by

$$nCO_2 (g) + nH_2O (l) \xrightarrow{h\nu} (CH_2O)_n + nO_2 (g)$$

In the presence of light, carbon dioxide and water are used to synthesis sugars and starches. In the higher plants, oxygen is released into the atmosphere. In this study we will be dealing with the somewhat simpler process observed in bacteria, as excellent crystallographic structures are now available to aid in these studies[1,2]. Bacterial photosynthesis does not release oxygen.

Much is known about the process of photosynthesis once the carbon monoxide is fixed[3-5], but controversy abounds in the initial photochemical event - the first fraction of a second after light is absorbed.

© 2004 American Chemical Society

And it is this initial event which is the study of this lecture.

Light is absorbed, generally by antenna pigments, and this energy is transferred to the reaction center (the "RC"), as shown schematically in Figure 1. This energy is then used by the reaction center to separate charge, from a Donor, D, to a chain of Acceptors, A. Essentially light energy is used to charge a capacitor, where this energy is now available to help break the strong bonds of CO_2.

It is now known that in the bacterial systems the donor is a pair of chlorophyll molecules, the "special pair", P, and the initial acceptors is either another chlorophyll molecule, an auxiliary chlorophyll B, or a bacteriopheophytin, H, a chlorophyll molecule in which the central magnesium ion has been replaced by a pair of protons. Although many pigments are present acting as antenna molecules to gather the light, additional chlorophylls are most common. How can the same molecule have such different roles? In spite of the fact that the antenna and the RC pair molecules are the same, light is transferred from the antenna to the RC with unit efficiency.

The heart of this system is chlorophyll, an example of which is given in Figure 2, where it is compared with the heme molecule, a porphyrin, ubiquitous in biology. Chlorophyll differs in several significant ways from porphyrins. Note that ring IV is saturated. The double bond has been broken. In the bacteriochlorophylls, ring II is also saturated. Is this important. What roll does this play, if any? All active chlorophylls also posses ring V, a five member ring with a carbonyl group. Why? Note also that all photosynthetic systems contain Mg as the central atom. Why is this? Certainly there is a lot more calcium, for example, in the environment. Is this important? Nature has had a long time to perfect a system for fixing carbon atoms, yet has chosen this one, and I will, throughout this lecture, make the assumption that this system has been highly selected to do its job - that all parts of the story are unique and important.

A schematic of the actual RC from the photosynthetic bacteria *Rhodopseudomonas viridis* [1] is given in Figure 3. Light is absorbed either directly by the special pair, P, or energy is transferred to the special pair from the antenna. Then within 3 picoseconds an electron is transferred from P to BL, the bacteriopheophytin associated with the L (Light) branch. This electron is than transferred to the quinone, QA in Figure 3, in about 200 ps, then across the iron atom to the quinone QB on the M-side in 100µs. Electrons flow from the cytochromes pictured above the membrane in Figure 3 to the pair allowing this process to be repeated an additional time. When there are two electron equivalents on QB, QB leaves its site and "docks" on the membrane. This "charged

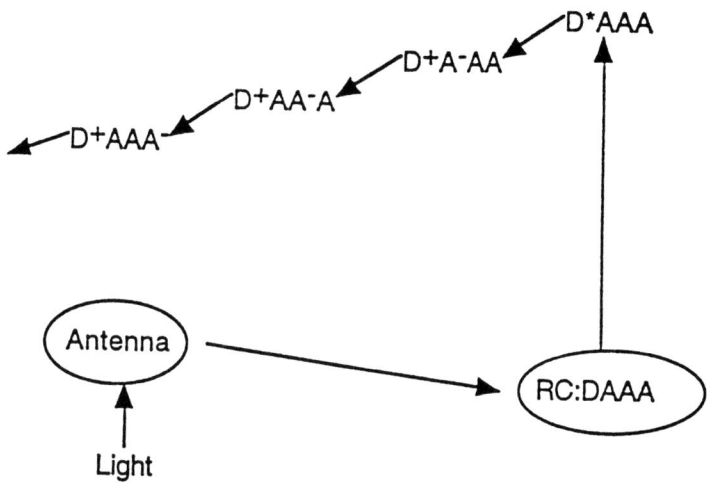

Figure 1: A Schematic of the absorption of light in a photosynthetic system. Light is absorbed initially by antenna pigments, and this energy is transferred into the reaction center, RC, where it is used to separate charge, with donor D giving up an electron to a series of acceptors A.

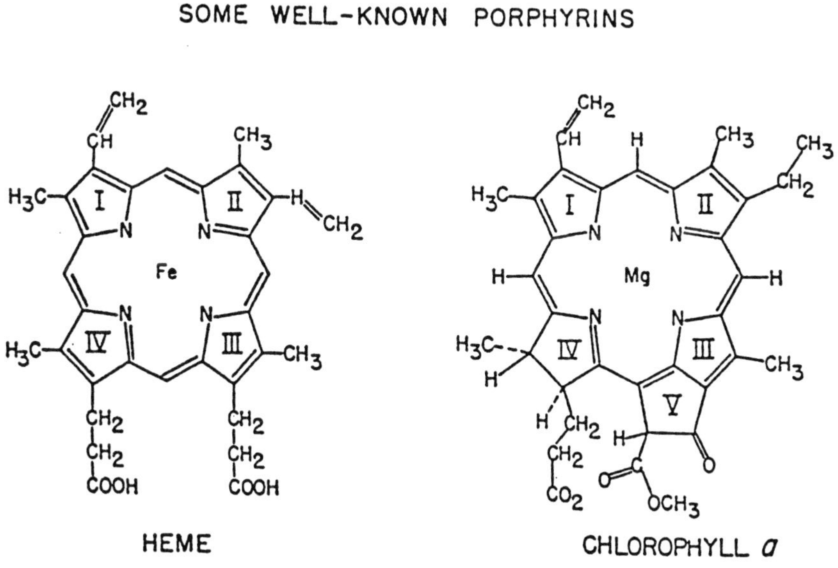

SOME WELL-KNOWN PORPHYRINS

HEME   CHLOROPHYLL *a*

Figure 2. A comparison of heme and chlorophyll a. Bacteriochlorophyll differ from chlorophyll in that the exterior double bond of ring II is saturated as it is in ring IV.

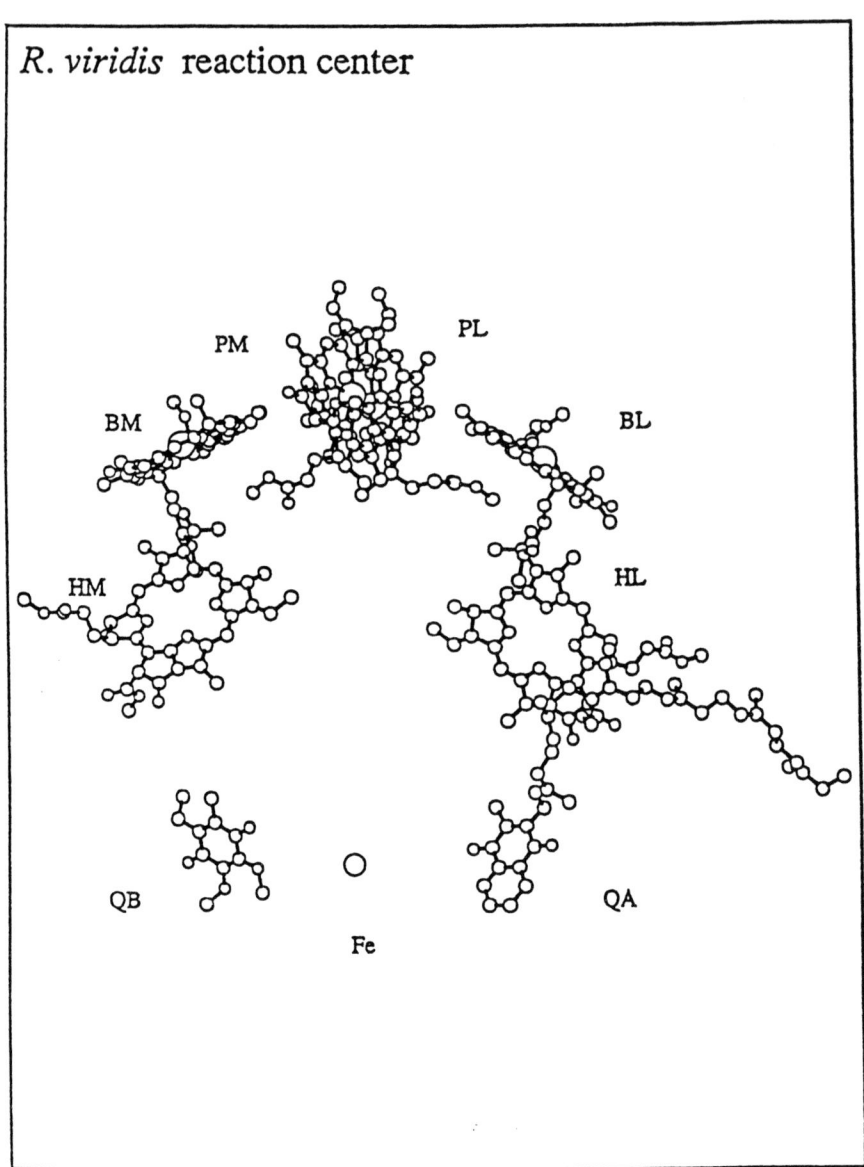

Figure 3: The structure of the reaction center of *r. viridis* taken from reference 1.

capacitor" than supplies the energy for the chemistry that follows[3-6] Evidence that the electron visits the auxiliary bacteriochlorophyll BL on its way to HL is controversial[7]. Is this reasonable? That the electron should jump over HL? And in spite of a near two-fold axis of symmetry between the L and M (Medium) sides, all of the electron flow is along the L side. Interesting questions abound in this process, and we catalog some of them in Figure 4.

For these studies we will use the molecular Schrödinger equation,

$$H\Psi_I(1,2,...) = E_I\Psi_I(1,2,...)$$

$$\Psi(1,2...) = A\,[\phi_1(1)\alpha(1)\ \phi_1(2)\beta(2)\ \phi_2(3)\alpha(3)\ \phi_2(4)\beta(4)\ ...]$$

to generate the properties of the ground and excited states of first porphyrin and chlorins, to determine the differences between them. We will then examine the dimer, to determine what special properties makes this structure unique. We will finish this article by considering the entire reaction center, representing a self-consistent field (SCF) configuration interaction (CI) calculation, in our largest case, of nearly 1000 atoms and 3000 electrons.

In the equations above, Ψ_I is the many electron wave function for state I with energy E_I, and ϕ_i is molecular orbital "i". When a specific electron label (j) is not given in the antisymmetrized product, $A[\cdots]$, lexicon order is assumed.

Figure 5 schematically shows the results of an SCF calculation on the ground state of a closed-shell molecule such as those we consider here. Excited states are then generated by removing one or more electrons from an occupied orbital (ϕ_i of the figure) and placing this electron into an empty orbital (ϕ_a). This generates a singlet and triplet wave function,

$$^{1,3}\Psi(1,2,...) = A\{[\phi_1\alpha\,\phi_i\beta\,...\,\phi_i\alpha\,\phi_a\beta\,...] \pm [\phi_1\alpha\,\phi_i\beta\,...\,\phi_a\alpha\,\phi_i\beta\,...]\}/\sqrt{2}$$

for which there are closed expressions for the excited states assuming that the molecular orbitals do not change[8],

$$^{1,3}\Delta E_{ia} = \varepsilon_a - \varepsilon_i - J_{ia} + (2,0)\,K_{ia}$$

In the above ε_a is the orbital energy of ϕ_a, ε_i that of ϕ_i, J_{ia} is the Coulomb repulsion between an electron in orbital ϕ_a and orbital ϕ_i and

1. Why magnesium? Is this atom special?
2. Why chlorins? Nature has broken the double bonds of ring IV, and, in bacteria, perhaps ring II. Is this significant?
3. Why is the presence of ring V important?
4. Is the "special pair" itself important? What "special" role does this play?
5. How does energy transfer take place? The energy transfer between the antenna chlorophylls and those of the reaction center occurs with unit efficiency. How is that possible as the molecules involved are identical?
6. There is an apparent symmetry between the L and M side yet all of the electron transfer is down the L side. Why does electron transfer not occur down the M branch?
7. What role does the Auxiliary BChl (B) play?
8. The iron atom of Figure 3 does not appear to play any role in the electron transfer process, although surely the electrostatic field a +2 ion generates is felt. Does it play a more specific role?
9. Does the protein play a direct role or does it simply provide scaffolding to hold the structure together?

Figure 4: Some of the questions that immediately arise when considering the initial photochemical event in photosynthesis.

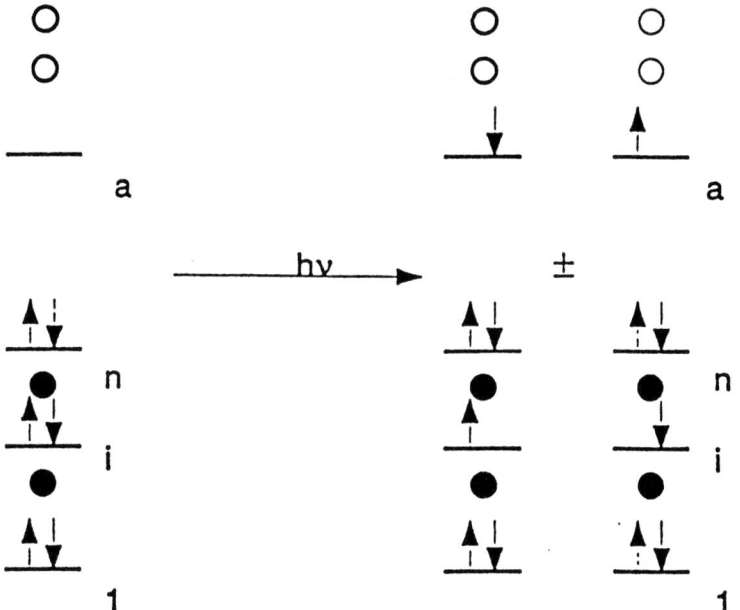

Figure 5: A schematic representation of the closed shell ground state Ψ_0 and an excited state generated by exciting an electron from mo ϕ_i to ϕ_a.

K_{ia} the corresponding exchange integral [9]. In the closed-shell ground state wave function we can identify ε_i and ε_a with the ionization potential and electron affinity, respectively, with removing an electron from ϕ_i and adding one to ϕ_a. To understand the frozen orbital expression for the excited state, we might identify ε_i with an ionization process, but then we cannot identify ε_a with the electron affinity of the neutral system, as the excited state has now but one electron in ϕ_i. In the neutral system there were two repulsions between the electron in ϕ_a and the two electrons in ϕ_i, and now there is but one. To correct this, we note that $\varepsilon_a - J_{ia}$ is the electron affinity for the electron in ϕ_a when there is only one electron in ϕ_i. There is no classical interpretation of the exchange term K_{ia}, but rather since K is always positive [10], this equation reflects Hund's rule, i.e., that states of highest multiplicity generally lie lower in energy than those of lower multiplicity[11]. We will return to this interpretation of the equations for the excitation energy to better understand the physical nature of the dimer interaction.

The above is only a rough outline of how excited states of molecules are estimated. In practice, configuration interaction is required, as the most appropriate orbitals to describe the excited states are not those of the ground state, and one must include the correlated motions of the electrons that are not included in the Hartree-Fock independent particle SCF procedure[9].

2. CALCULATIONS ON MONOMERS:

The spectroscopy of porphyrins has a long and eloquent history, including the work of Simpson[12], Kuhn[13], Platt[14], and Gouterman[15]. Part of the fascination no doubt derives from the relatively weak visible Q band which makes these compounds very colorful, and the unusually intense Soret, or B band, that has extinction coefficients stretching to over a million[16]. A summary of the results of our calculations is given in Figure 6 [17]. The important molecular orbitals, shown at the top of this figure, are the frontier occupied orbitals $a_{1u}(\pi)$ and $a_{2u}(\pi)$, and the lowest empty orbitals, a pair of degenerate orbitals labeled $e_g(\pi)$. The four excitations among these orbitals generate two nearly degenerate strongly allowed excitations of ^1E_u type. These two states undergo configurational mixing, forming nearly equal mixtures of one another. The transition dipoles of the state depressed in energy nearly cancel, giving

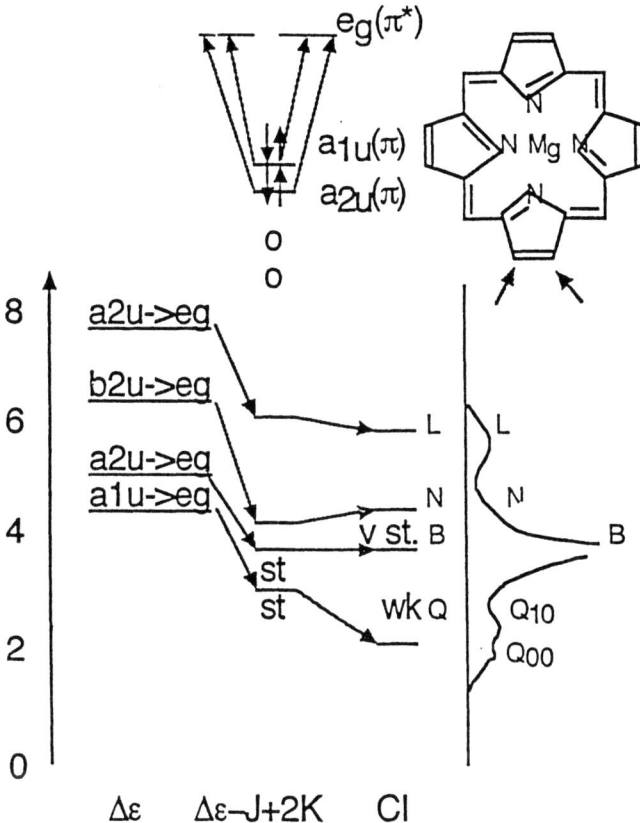

Figure 6: The results of calculations on Mg porphin. The active molecular orbitals are shown in the upper part of the figure. $\Delta\varepsilon$ refers to the orbital energy difference, and $\Delta\varepsilon - J + 2K$ to the frozen orbital approximation for the excited state energies. The third column represents the results after CI. st = strong, v.st = very strong, wk = weakly allowed, N and L are the labels of the states observed at higher energy than the B (Soret). The group theory labels used are those of D_{4h}. From ref. 17.

rise to the weak visible Q band. In the combination that is elevated in energy through this mixing the transition dipoles reinforce, giving rise to the very strongly allowed Soret, or B band. As the figure indicates this is somewhat of an oversimplification. Other higher lying configurations in the CI treatment also interact, especially with the higher energy B band, depressing it in energy, and even stealing some of its intensity[18]. Nevertheless, the four orbital *concept* describes the situation very well, even if it does not reproduce some of the details quite correctly. In general, the calculations we describe reproduce the nature of the Q band quite nicely, but B bands are calculated too high in energy, typically by 4000 cm^{-1} to 5,000 cm^{-1} [17,19,20,21].

Although porphyrins are quite colorful, they are not particularly good choices for photosynthetic pigments. Nature would prefer, given enough time, to choose a species which absorbed more highly in the visible, rather than in the near UV.

Figure 7 summarizes our results for the Mg chlorin[17]. Here the four fold axis has been destroyed along with one of the exterior double bonds of the pyrrole ring. The near degeneracy of the occupied $a_{1u}(\pi)$ and $a_{2u}(\pi)$ orbitals present in porphyrins is destroyed, as is the degeneracy of the $e_g(\pi)$ unoccupied orbitals. Although the four-orbital model still leads to two strong states, they are no longer nearly degenerate, and they mix only very slightly upon configuration interaction. Thus both the Q and the B band are predicted to be strong, and this story does not change when two exterior pyrrole double bonds are saturated. Thus one might conclude that mother nature has chosen to break these double bonds of porphyrin to throw intensity into the normally weak Q (visible) band to better absorb in the visible. After all, even the most efficient electron transfer system will not be useful in photosynthesis if it did not strongly absorb the available light!

The absorption spectra of Mg etioporphyrin and Chlorophyll a are contrasted in Figure 8.

3. CALCULATIONS ON THE SPECIAL PAIR.

Calculations on porphyrins have been done now many times using many different techniques. Those reported in figures 6 and 7 are done using the semi-empirical INDO/s model [22-24], a scheme that has been parametrized directly on molecular electronic spectra at the CI - singles level. All of these methods essentially give the same or similar results, although few are as precise as is the INDO/s scheme for actually

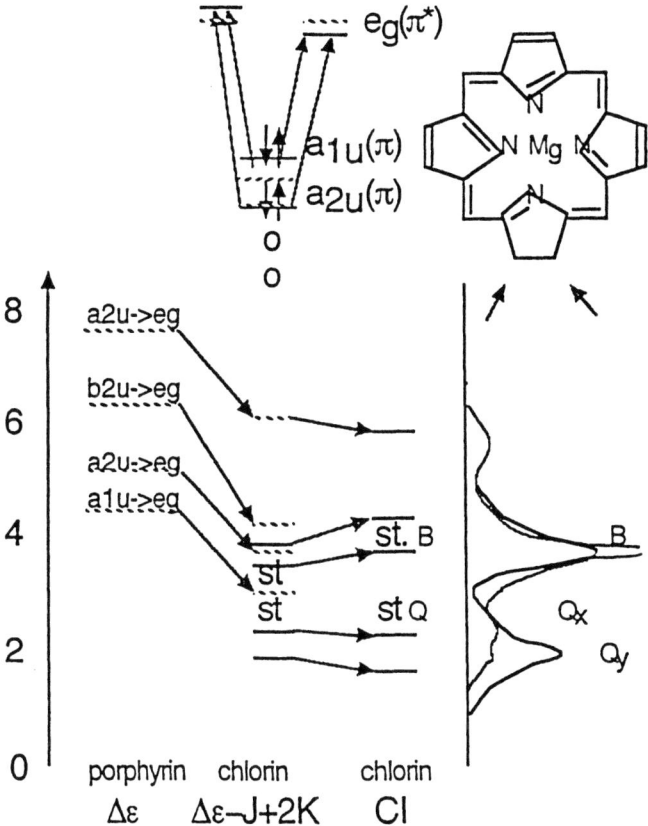

Figure 7: The results of calculations on Mg chlorin, contrasted with those of Mg porphin, see figure 6. The greater splitting of the two low lying strong bands before configurational mixing, lead to considerably less mixing in the CI, and thus both Q and B bands are predicted to be strong, after ref. 17.

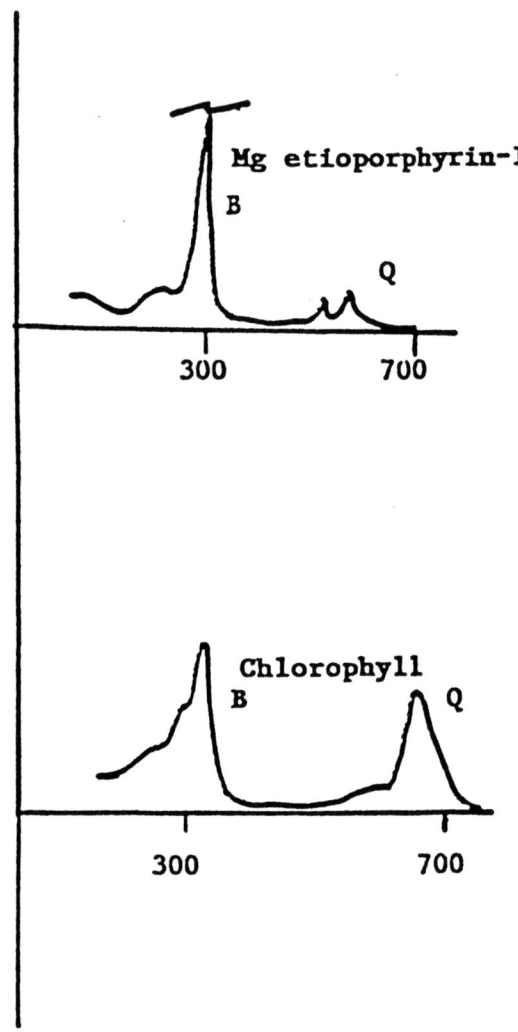

Figure 8. A comparison of the spectra of Mg etioporphyrin and chlorophyll a.

predicting transition energies so close to those observed.

The calculations we now report on the larger systems are exclusively of the INDO/s type. Following the flow chart of Figure 9, we now pursue the nature of the excited states of the special pair. The special pair that we have examined is show in Figure 10 [1, 19]. The results that we obtained for our most complete calculation are given in Table I. These results are in very good agreement, indeed, with the observed band at 10,300 cm^{-1} [15]. Although the agreement with experiment is very satisfying, it is fair to say that it was difficult to give physical significance to these results. To learn more about this system, we modeled it with the idealized system shown in Figure 11. The presence of the inversion center allows us to resolve the excited states easily in terms of the monomeric units[19]. We then studied the calculated spectroscopy of this

Table I: Calculations on the Qy1 (P*) band of the special pair in *R. viridis*. Energy is given in cm.$^{-1}$. f refers to the calculated oscillator strength, and TX to M to L side charge transfer in this excited state.

system	ΔE	f	TX
1. Model dimer	12,400	1.19	0.00
2. Native dimer	10,793	0.99	0.16
3. Case 2 with histidines	11,338*	1.01*	0.12*
4. Case 3 plus H-bonding three amino-acids (L168, L248 and M195)	10,002	1.00	0.26

* smaller CI with only 197 detors.

idealized dimer as both a function of inter-ring distance and twist angle [19,26]. Before we do so, however, I wish to briefly examine the coupled chromophore model, briefly outlined in Figure 12. Although couple chromophore theory generally predicts a doubling of states, there is, in fact, a quadrulping of states that needs to be considered in this case. The four-orbital model predicts four states for excitations between a1 and a2 and a3 and a4, giving rise to Qx, Qy and Bx and By transitions. In the dimer, however, there are four occupied and four empty orbitals to consider, giving rise now to 16 states. In general, half of these states are of high energy, as they involve charge transfer from one monomer unit to the other. In this case, however, as we shall see, these charge transfer states are low in energy and contribute in a significant way to the observed transition energies of the lowest lying states.

Figure 13 presents a stick diagram of the results we obtained for the model dimer with an inter-ring separation of 5 Ångstroms (the observed

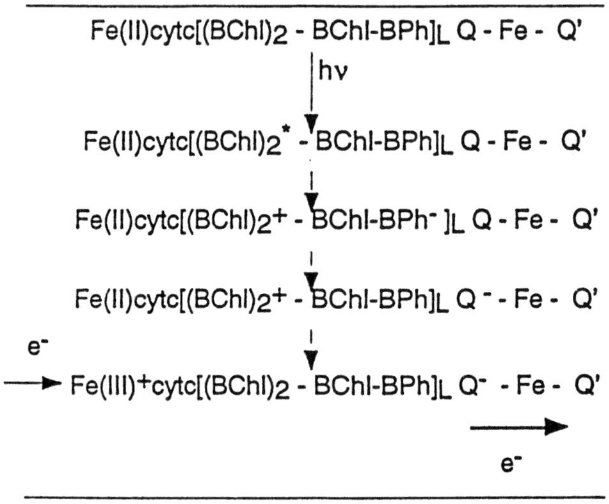

20

$$Fe(II)cytc[(BChl)_2 - BChl\text{-}BPh]_L\ Q - Fe - Q'$$
$$\downarrow h\nu$$
$$Fe(II)cytc[(BChl)_2^* - BChl\text{-}BPh]_L\ Q - Fe - Q'$$
$$\downarrow$$
$$Fe(II)cytc[(BChl)_2^+ - BChl\text{-}BPh^-]_L\ Q - Fe - Q'$$
$$\downarrow$$
$$Fe(II)cytc[(BChl)_2^+ - BChl\text{-}BPh]_L\ Q^- - Fe - Q'$$
$$\xrightarrow{e^-} Fe(III)^+cytc[(BChl)_2 - BChl\text{-}BPh]_L\ Q^- - Fe - Q'$$
$$\xrightarrow{e^-}$$

Figure 9. A summary of the events shown in Figure 3.

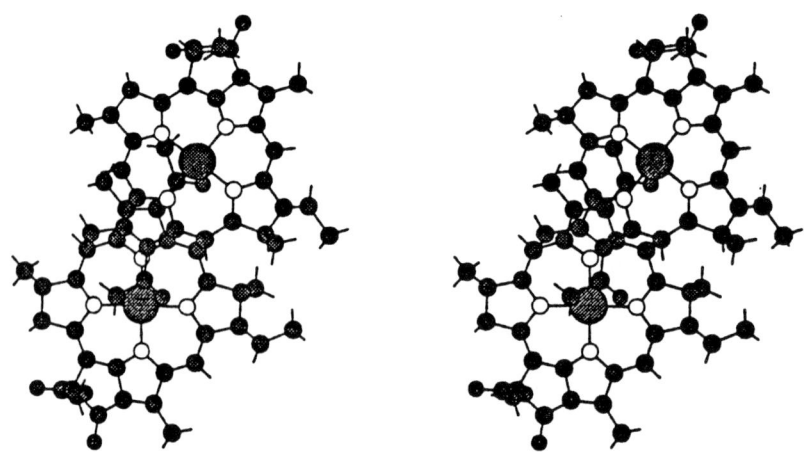

Figure 10. Stereo view of the special pair of *R. viridis*. See Table I.

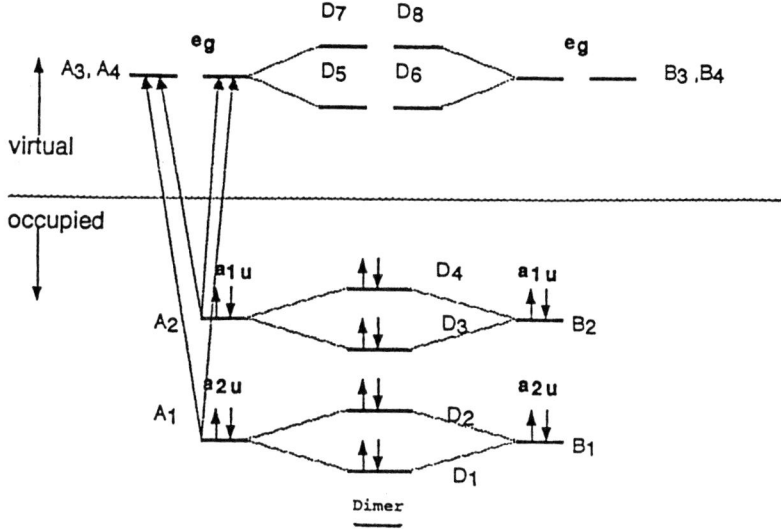

Monomer A   Monomer B

Figure 11. The idealized dimer modeled in these calculations.

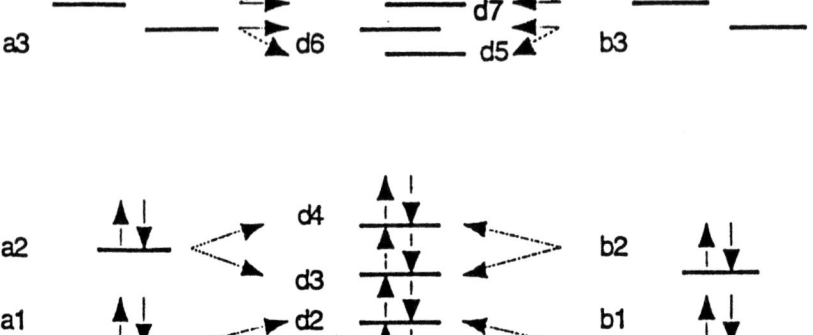

Figure 12. A representation of the coupled chromophore model. The orbitals designated (a) represent the frontier four-orbitals of one monomer chlorophyll unit, those labeled (b), the orbitals of the second monomer unit. Upon mixing, symmetry determines ± combinations of the monomer orbitals, yielding those of the dimer, labeled with (d).

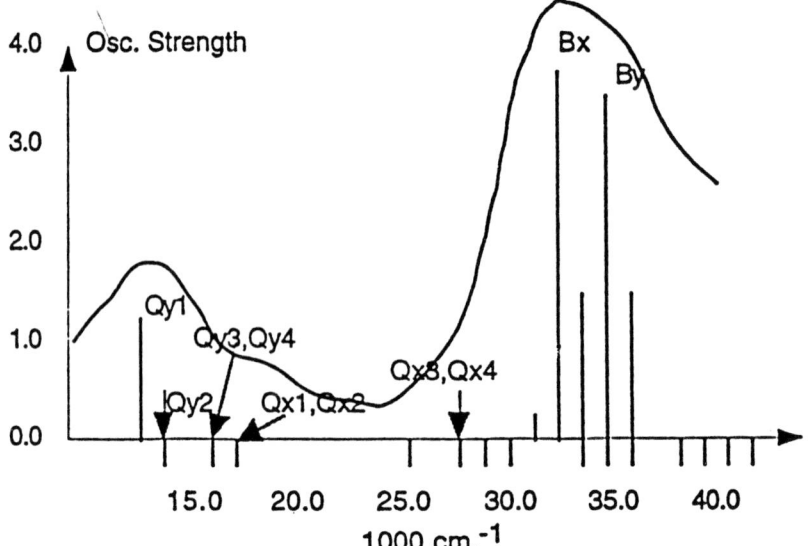

Figure 13: The calculated spectrum of the model dimer of figure 10 at 5 Å. Lines drawn with negative oscillator strength indicate the positions of transitions with no calculated oscillator strength. Y is the long axis of figure 10.

separation is about 3.3 Å). The four Qy bands lie lowest in energy, and these are the photochemically most active transitions, also presented in Table II. Although the model system, even at 5Å separation, seems to be acting as a true super-molecule, we can resolve the excitations by writing each of the dimer orbitals as

$$d_i = (a_i + b_i)/\sqrt{2}$$
$$d_{i+1} = (a_i - b_i)/\sqrt{2}$$

and expanding the determinants given in Table II. This leads to a description of the excited states as excitonic or charge resonance, see Figure 14.

Table II. The nature of the calculated Qy transitions in the model dimer at 5 Å. CR designates a charge resonance state, see text. The orbitals designated d_i refer to the dimer orbitals of Figure 12.

State	Qy1	Qy2	Qy3	Qy4
Energy	13,081	13,175	16,936	16,937
Osc. Str.	1.263	0.000	0.000	0.000
Excitation	Exciton	Exciton	CR	CR
d4 -> d5	0.6711	0	0	0.7028
d3 -> d6	-0.6617	0	0	0.7114
d4 -> d6	0	-0.6656	-0.7063	0
d3 -> d5	0	0.6639	-0.7079	0
d2 -> d7	-0.2376	0	0.0019	0
d1 -> d8	-0.2351	0	0.0019	0
d2 -> d8	0	0.2419	-0.0004	0
d1 -> d7	0	0.2401	-0.0004	0

We discuss here the nature of Figure 13 in more detail. As mentioned previously, we can estimate excitation energies using the equation $\Delta E = \varepsilon(e_g) - \varepsilon(a_{1u}) - J(e_g, a_{1u}) - 2K(e_g, a_{1u})$, where the mo labels now refer to porphyrin for convenience. In the case of excitonic excitation, the quantities in this equation are essentially those of the monomeric units. In the case of charge transfer, $\varepsilon(a_{1u})$ does refer to an ionization process, as we are removing an electron from one of the monomeric units, and $\varepsilon(e_g)$ refers to an electron affinity, as we are adding an electron to the

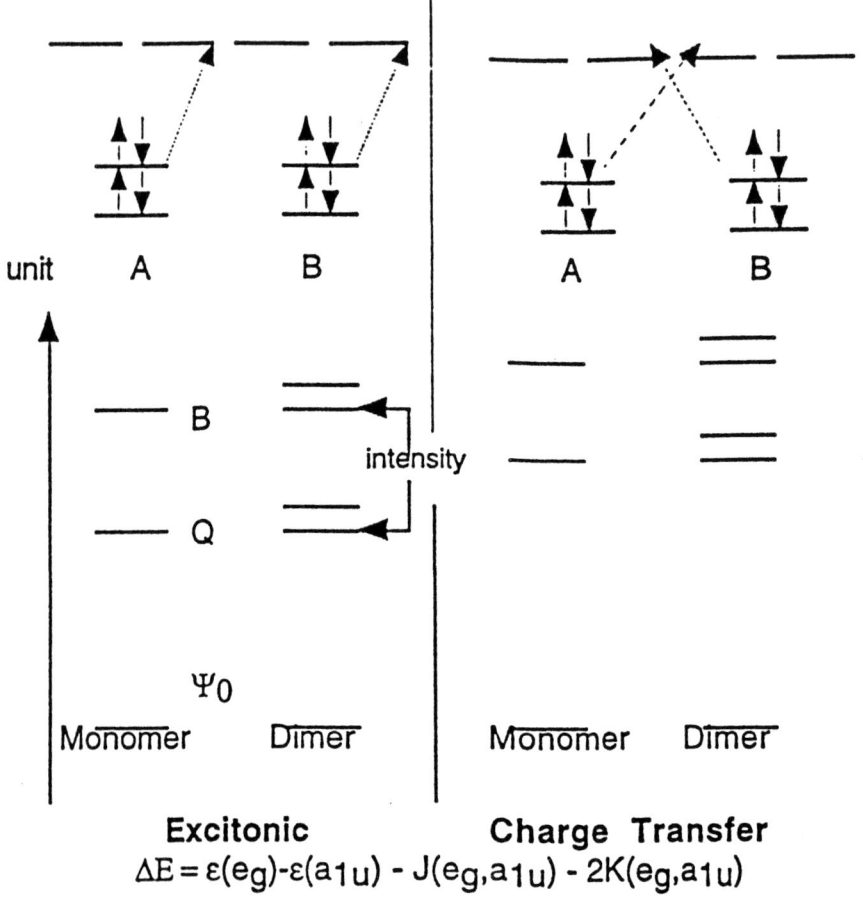

Figure 14. The interaction of two porphyrin-like units to yield a dimer. The left side of the diagram shows the interaction of two local excitations, yielding a doubling of states and the B and Q bands. The oscillator strength lies in the lower combination of localized excitations, see text. The right side of the diagram shows the coupling of two charge transfer excitations to yield the two charge resonance excitations, generally at higher energy than the excitonic states.

other. This difference, $\varepsilon(e_g)-\varepsilon(a_{iu}) \sim$ EA - IP, is reduced by the Coulomb attraction term between the positive unit and the negative one, at a distance ~R, or $J(e_g,a_{1u}) \sim 1/R$. For this reason we expect the position of the charge resonance transitions to be sensitive to inter-ring separation.

Table III. Resolution of the model dimer states into excitonic and charge resonance components. Energy in 1000 cm.$^{-1}$

		---------------- 5Å ----------------------				------------------- 3.5Å ------------		
		Ex	Ex	CR	CR	Ex	Ex	CR	CR
ΔE		13.0	13.2	16.9	16.9	12.4	13.4	15.0	15.5
f(osc)		1.26	0.00	0.00	0.00	1.19	0.00	0.00	0.10
L	A	0.67	0.66	-0.00	-0.00	-0.62	-0.66	-0.04	0.25
O	B	-0.67	0.66	-0.00	0.00	0.62	-0.66	-0.04	-0.25
C	A	-0.21	-0.24	0.00	0.00	-0.21	-0.25	-0.02	0.09
A	B	0.23	-0.24	0.00	0.00	0.21	-0.25	-0.02	-0.09
L									
C	A>B	-0.00	0.00	0.71	-0.71	0.26	-0.04	0.71	-0.66
R	B>A		0.00	0.00	0.71	0.71	-0.26	-0.04	0.71
		0.66	O A>B	0.00	-0.00	0.00	0.00	0.01	-0.01
		-0.00	-0.01						
S	B>A	0.00	-0.00	0.00	0.00	-0.01	-0.01	-0.00	0.01
S									

Table III shows that at 5Å separation, the excitations nicely classify as either excitonic or charge resonance (the combination of two charge transfer excitations that exactly cancel any net charge transfer in the state.). As shown in Table III, this is not the case for the system at 3.5 Å. We note that at the 5Å separation there is a small splitting of 300 cm.$^{-1}$ of the two excitonic bands, and no calculated splitting of the two charge resonance bands calculated some 3,700 cm.$^{-1}$ higher in energy. From the CI coefficients after projection the two lowest bands, Qy1 and Qy2, are pure excitonic, the two higher, Qy3 and Qy4, pure charge resonance.

At 3.5Å the splittings are considerably larger, as one might expect from the closer distance. But the splitting is larger than expected because the two CR transitions have come down in energy, a consequence of the larger 1/R term discussed previously. The lower CR excitation, Qy3, is now only 1,600 cm.$^{-1}$ higher in energy than the upper excitonic transition, Qy2. And it is clear that Qy1 has picked up charge resonance character (see the cross term coefficients in Table III) and Qy4 has picked up excitonic character (see the enhanced oscillator strength and local coefficients in this table). In addition to the excitonic splitting, there is an extra interaction between Qy1 and Qy4, that further lowers Qy1.

We believe this lowering is significant [20,27], as it guarantees that the lowest excited state in the system belongs to the special pair. Recall that the energy associated with the light absorbed by the chlorophyll antenna molecules is transferred with unit efficiency to the reaction center. The forming of this dimer guarantees through this interaction that the lowest excited state of all the chlorophylls present will be associated with the pair. It is tempting to speculate that this pair will be present in all photosynthetic systems as it is such a robust way of insuring that the lowest excited state of the system will belong to the RC.

4. CALCULATIONS ON THE ENTIRE RC.

We now turn our attention to the entire reaction center, and report on some of the calculations we have performed. The largest of these calculations involves nearly 1000 atoms and 3000 electrons, although those we report below are smaller[28].

Considering some 600 atoms and 1200 electrons of the reaction center, essentially all 6 chromophores and the chelating histidines, we obtain the results as shown in Figure 15. Qy1, or P*, is calculated at 11,600 cm.$^{-1}$, higher than the 10,002 cm.$^{-1}$ reported in Table I [20], a consequence of the fact that we cannot include as many configurations localized to the pair in this much larger calculation than we did when considering only the pair. As anticipated, P* is the lowest calculated excited state, *after* errors in the crystallographic coordinates were corrected. Somewhat disappointing at first glance is the relatively high transition energies for the P to B or P to H electron transfer transitions, all calculated at about 20,000 cm.$^{-1}$, about 1eV, or 23 Kcal/mol. higher in energy than P*. It is difficult to envision a vibrational degree of freedom sufficiently large enough to allow crossing from the P* state into any of the charge transfer states if they lie so much higher in energy.

The center-to-center distance between P and H is 17 Å, and moving an electron this distance might be expected to generate a dipole moment of some 80 Debye. It is perhaps naive to expect the polarizable material that surrounds the chromophores not to respond to such a large separation of charge.

To examine this, we make use of the self-consistent reaction field model [29-31] that treats the surrounding polarizable matter as a dielectric continuum. This is, of course, a very simple approach, but it is not possible to actually include enough of the protein and its surroundings to model this effect directly.

In this model we allow the charge distribution of the RC to stabilize with respect to the surroundings using a model that goes back to Born [32], Kirkwood[33], Onsager[34] and Lippert [35]. In this model the total energy of the system can be written, after a multipole moment expansion of the potential, as

$$E(\text{univ.}) = E(\text{gas}) - 1/2 \; \Sigma_{(l)} \; g_1(\varepsilon) \Sigma_{(m)} \; M_{lm} \; M_{lm}$$

in which M_{lm} are the moments, i.e., the monopole (charge), the three components of the dipole, etc., and g_1 is the reaction field tensor. The first two terms of this expansion, for example, are the familiar Born charge term

$$e_0 = -1/2 \; (\varepsilon-1)/ \; (\varepsilon a_o) \; q^2$$

and Onsager dipole term

$$e_1 = -1/2 \; (2(\varepsilon-1))/ \; ((2\varepsilon+ 1) \; a_o^3) \; \mu \cdot \mu$$

where ε is the bulk dielectric constant, a_o is a cavity radius(that we have chosen from mass density), q the net charge of the system, and μ the dipole moment.

Using the variational principle, we derive [29,31]

$$f \; \phi_i = \varepsilon_i \phi_i$$

with f the Fock operator in the usual fashion. However, in this case the Fock operator is modified to reflect the presence of the dielectric material, which is given by

$$f = f_o - \; (2(\varepsilon-1))/ \; ((2\varepsilon+ 1) \; a_o^3) <\mu> \cdot \mu$$

where we demonstrate only the dipolar term for convenience. The dipolar term is also shown schematically in Figure 16 [36].

We assume that the solute molecule, the RC in this case, is in equilibrium with its surroundings. Upon absorption, a very fast process, we assume that there is not enough time for the nuclear degrees of freedom of the solvent (protein) to respond. Rather we further assume that only the electronic polarization of the surrounding media responds. This is shown on the left hand side of Figure 16 by a partial rotation of the solvent's induced dipole [31-37].

In Table IV we report our results [20]. As expected the states that have the largest dipole moments are lowered in energy more than are the states with a smaller asymmetry of charge. Quite remarkably the state P^+HL^- is now predicted to lie at 12,600 cm.$^{-1}$, only 1,200 cm.$^{-1}$ above P^*

Table IV: The calculated Spectroscopy of the Reaction Center from *R. viridis*. Energy in 1000 cm.$^{-1}$ and dipole moment differences $\Delta\mu$ in Debye.

State	"Gas phase"		Solvent simulated	
	Energy	$\Delta\mu$	Energy	$\Delta\mu$
$Qy1 = P^*$	11.6	4.6	11.4	4.6
P^+HL^-	19.7	78.9	12.6	75.0
P^+BL^-	19.8	49.9	16.6	46.1
P^+BM^-	20.2	55.5	16.7	48.4
P^+HM^-	20.6	76.9	13.5	74.8

There are many possible ways now for the energy to cross from P^* into P^+HL^-, including further dielectric relaxation. Table IV also shows that the state P^+BL^- is not lowered nearly as much, as the relaxation term is roughly proportional to the change in the dipole moment squared. These calculations do not support a direct role of this state in the charge transfer process, although we can not rule out the participation of this state in super-exchange [38,39].

It is also interesting to note that all excitations on the M-side are calculated higher in energy than those of the L-side. Although this does not address the dynamics of the process, it does already point out a thermodynamic preference of L-side electron transfer over the M-side of 60:1 at room temperature, using the Bolzmann equation and the calculated 860 cm.$^{-1}$ difference between P^+HL^- and P^+BL^-, and that there is enough asymmetry in the coordinates of the reaction center to cause such differences: i.e., that the structure does not have two fold symmetry. Others have addressed the dynamics of the asymmetry in the electron transfer[39-41].

We summarize some of these findings in Figure 17.

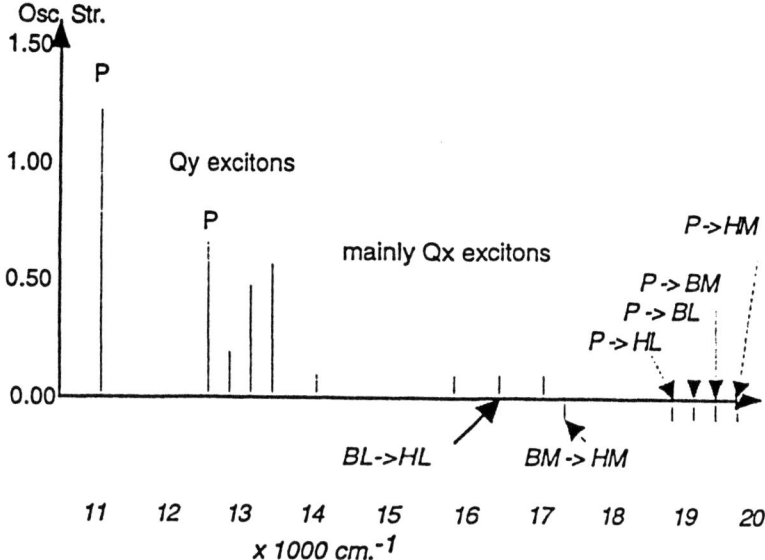

Figure 15: The calculated spectrum of the RC, taken from reference 20.

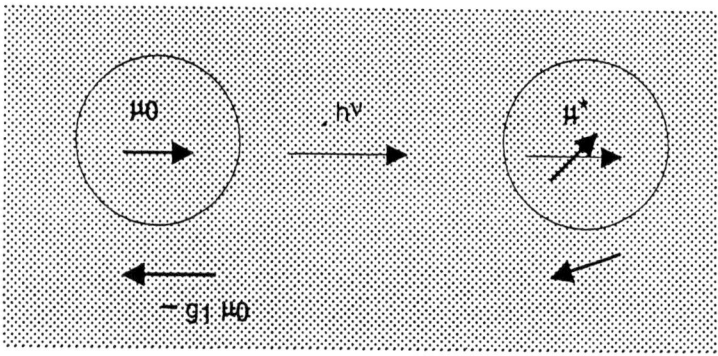

Figure 16: This figure shows the stabilization of a molecule through the moments induced in a polarizable material do to the charge density of the solute. Only the dipolar term is shown. Upon excitation we assume that the nuclear degrees of freedom are too slow to change and affect the absorption frequency, but that the electrons of the solvent do respond, see text.

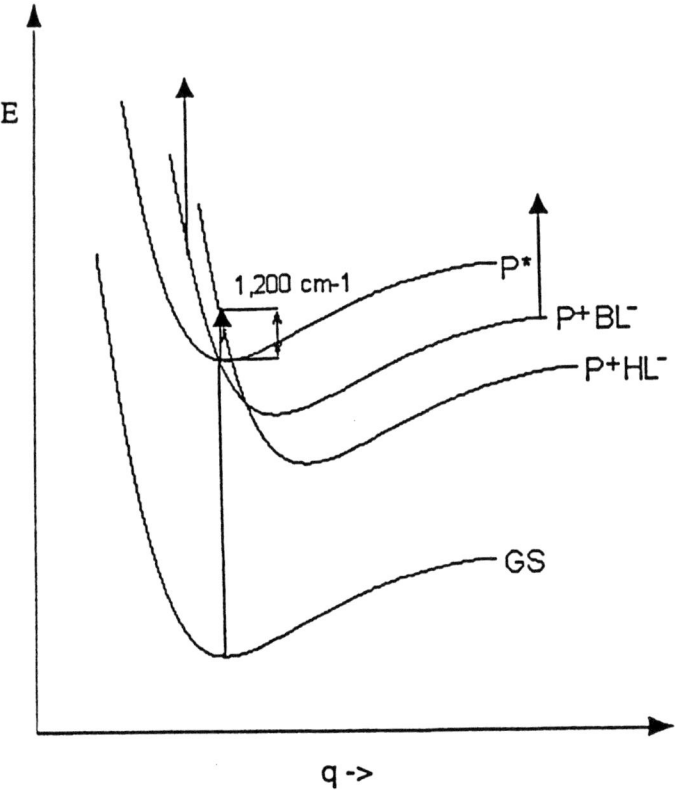

Figure 17. The results of these calculations suggest that excitation from the ground state is directly into P*, and that the system can cross directly into the state P+HL- upon further nuclear relaxation of the protein, or of the positions of the RC chromophores relative to one another. The state P+BL- represented in dash line, is placed where it should appear if it were to play an active role in the charge transfer process as an intermediate. These calculations place this state too high in energy to actually be populated during the electron transfer.

V. CONCLUSIONS:

We conclude with a reexamination of the questions posed in Figure 4.

1. Why Mg? No doubt this has to do with the exact positioning of P* with respect to the charge transfer states, especially P^+HL^-. Replacing Mg with Ca changes the calculated absorption of the pair significantly, but this does not seem to be the case for Zn. At the level of resolution of these calculations we cannot rule out a Zn based photosynthetic system!

2. Why the chlorins? We conclude that by breaking the double bonds in the pyrrole rings and destroying the 4 fold symmetry of the porphyrins intensity is transferred from the very intense porphyrin B band to the lower energy Q band, making the photosynthetic pigments much more absorbent in the visible, and much better able to use the available light than would be the porphyrins

3. Why ring V, the five member ring with the carbonyl group? It has been suggested in the past that this ring is involved in resonance structures that lower the ionization potential of the chlorophylls relative to those of simple chlorins or porphyrins. We have examined this hypothesis, and find that Mg chlorins, simple chlorophylls and porphyrins all have calculated gas phase ionization potentials all the same, about 6.2 eV to 6.4 eV[17]. We have also examined the role that ring V plays in the calculated transition dipole in the special pair, and find nearly the same magnitude and direction for the pair with or without ring V[42]. Ring V likely plays an important role in the structure of the reaction center, and, indeed, is involved in hydrogen bonding in the observed structures.

4. Why the dimer or special pair? We believe the formation of the dimer insures that the lowest lying excited state lies in the RC, aiding in the efficient energy transfer between antenna chlorophylls and the pair. Rather interesting is the interaction between the first excitonic state and the last CR state of the Qy manifold that leads to an "anomalous" red shift in the former.

It is also interesting to remark that the slipped structure of the dimer insures that it is the lower excitonic transition that has the calculated intensity, rather than the upper. A face-to-face arrangement of the chlorophyll molecules in the dimer would lead to the upper component having the intensity, and again, a less efficient energy transfer.

5. How does the energy transfer take place. We suggest that the formation of the dimer and the resultant low lying P* is part of the reason. Certainly utilizing chlorophyll molecules for both the antenna and the RC guarantees a good overlap of transition energies, a necessary condition for good energy transfer.

6. Why not the M branch? Again we have not done the dynamics, but

we do note from these calculations that there is enough asymmetry in the structure to yield excitations in the L-side all lower than those on the M-side. Our best calculations to date place P^+HL^- some 860 cm.$^{-1}$ below the P^+HM^-. Others that have examined the dynamics all suggest that the L-side is more effective in leading electron transfer, although there is not general agreement on the role of P^+B^- [38-41]. What role does the M branch than play? Did it have some role in the past, now obsolete? Is it only to hold the structure together in its rather unique geometry? Does it add a redundancy to the system that can take over in case of accident?

7. What role does the auxiliary bacteriochlorophyll play? We do not believe it plays a direct role in the electron transfer. Does it act to help electron transfer from P* to P^+HL^- through super-exchange? This should then appear as a CI coefficient in the CI wave function, and we find this mixing very small.

An examination of the structure shows that BL and BM extend beyond the L and M proteins where they might have good contact with the antenna pigments. Since they have excitation energies higher that P* and lower than the pigments it is tempting to involve them in the energy transfer process as bridges between the pair and the antenna.

8. What role does the non-heme iron play? This we have not examined - yet- in any detail. One role is certainly structural - it ties the L and M branches together through the chelating ligands, and thus helps to hold the chromophores together in the rather curious geometry that is observed. But what about the electrostatic field it creates? Does this change things? If this is all, than other +2 ions could also do the job?

9. What role does the protein play? One is certainly to hold the structure together. Another is that it supplies a polarizable environment important in stabilizing the charge transfer states, and allowing the intersystem crossing necessary for charge separation, the heart of this process.

The quantum chemical calculations that we have performed on these systems are approximate. We cannot necessarily demand great accuracy from semi-empirical quantum chemistry in general, or *ab-initio* quantum chemistry specifically, on systems of this size. I hope, however, to have convinced the reader that you can have <u>fun</u> with such calculations, and that you can get important ideas and suggestions that were not possible *without* these calculations. Likely the best results are those that we could say, after the calculations have pointed the way, were not really needed. The underlining physics of the situation are so appealing that they rest on no specific calculation at all

VI. ACKNOWLEDGEMENTS

I have had many important co-workers in this work, who not only supplied the actual calculations but many of the ideas. Certainly Dan Edwards (Idaho) who did much of the work on the monomers, and some of the dimer work. Mati Karelson (Tartu) was important in helping to develop the reaction field model for spectroscopy. Mark Thompson (PNL) did all the calculations on the RC that are discussed in this manuscript, and Marshall Cory (QTP, Florida) is carrying on this tradition.

Financial aid throughout this work has come from the Office of Naval Research and the National Science Foundation (CHE9312651) as well as from the University of Florida Division of Sponsored Research.

References.

[1] J. Deisenhofer, O. Epp, K. Miki, R. Huber and H. Michel, J. Mol. Biol. (1984) 180, 385; *ibid,* Nature (London, 1985) 318, 618; H. Michel, O. Epp and J. Deisenhofer, J. EMBO J. (1986) 5, 2445; J. Deisenhofer and H. Michel in *The Photosynthetic Bacterial Reaction Center Structure and Dynamics,* eds. J. Bretton and A. Vermeglio, NATO ASI Series 149, Plenum Press, New York (1988); J. Deisenhofer and H. Michel, Science (1989) 245, 1463.
[2] C. H. Chang, D. Tiede, J. Tang, U. Smith J. R. Norris and M. Shiffer, FEBS Lett. (1986) 205, 82.
[3] See *The Photosynthetic Bacteria*, ed. R. K. Clayton and W.R. Sistrom, Plenum Press, New York (1978).
[4] R.K. Clayton, *Photosynthesis: Physical Mechanisms and Chemical Patterns* , Cambridge Press, England, (1980) .
[5] J. Amesz, Photosynthesis, Elsevier, Amsterdam (1987).
[6] M. D. Hatch and N. K. Boardman in *The Biochemistry of Plants: Photosynthesis*, Vol. 8, ed. P. K. Stumpf and E. E. Conn, Academic Press, New York (1988).
[7] See, for example, W. Holzapfel, U. Finkele, W. Kaiser, D. Oesterhelt, H. Scheer, H. U. Stilz and W. Zinth, Proc. Natl. Acad. Sci. USA (1990) 87, 5168.
[8] R. G. Parr, *Quantum Theory of Molecular Electronic Structure,* Benjamin, New York (1963).
[9] A. Szabo and N. Ostlund, *Modern Quantum Chemistry,* McGraw Hill, New York (1989).
[10] J. C. Slater, *Quantum Theory of Atomic Structure*, Vol. 1. McGraw Hill, New York (1960).
[11]. G. Herzberg, *Spectra of Diatomic Molecules*, D. Van Nostrade, Princeton, New Jersey (1961).

[12] W. J. Simpson, J. Chem. Phys. (1949) 17, 1218.
[13] H. Kuhn, J. Chem. Phys. (1949) 17, 1198.
[14] J. R. Platt in *Radiation Biology*, vol. 3, ed. A. Hollander, McGraw Hill, New York (1956).
[15] M. J. Gouterman, J. Mol. Spectrosc. (1961) 6, 138.
[16] L. Edwards, D. H. Dolphin and M. J. Gouterman, J. Mol. Spectrosc. (1970) 35, 90.
[17] W. D. Edwards and M. C. Zerner, Intern. J. Quantum Chem. (1983) XXIII, 1407; W. D. Edwards, J. D. Head and M. C. Zerner, J. Am. Chem. Soc. (1982) 104, 5833.
[18] see, for examples, J. D. Petke, G. M. Maggiora, L. L. Shipman and R. E. Christoffersen, J. Mol. Spectrosc. (1978) 71, 64; *ibid*, 73, 311; *ibid*, J. Am. Chem. Soc. (1977) 99, 7470; *ibid*, 99, 7478.
[19] M. A. Thompson, M. C. Zerner and J. Fajer, J. Phys. Chem. (1991) 95, 5693.
[20] M. A. Thompson and M. C. Zerner, J. Am. Chem. Soc. (1991), 113, 8210.
[21] W. D. Edwards, B. Weiner and M. C. Zerner, J. Am. Chem. Soc. (1986) 108, 2196.
[22] M. Lösche, G. Feher and M. Y. Okamura in *The Photosynthetic Reaction Center Structure and Dynamics*, ed. J. Breton and A. Vermeglio, NATO ASI Series 149, Plenum Press, New York (1988) pp 151-164.
[23] J. Ridley and M. C. Zerner, Theoret. chim. Acta (Berlin) (1973), 32, 111; *ibid*, (1976) 42, 223.
[24] M. C. Zerner, G. Loew, R. Frischner and U.-T. Mueller-Westerhoff, J. Am. Chem. Soc. (1980) 102, 589.
[25] J. Breton, J. Biochim. Biophys. Acta (1985), 810, 235; J. Breton in *The Photosynthetic Reaction Center Structure and Dynamics*, ed. J. Breton and A. Vermeglio, NATO ASI Series 149, Plenum Press, New York (1988) pp 56-69.
[26] M. Thompson, PhD. Thesis. Dept. of Chemistry, University of Florida, Gainesville, Fl 32611 (1990).
[27] M. A. Thompson and M. C. Zerner, J. Am. Chem. Soc. (1988) 110, 606.
[28] M. J. Cory and M. C. Zerner, work in progress.
[29] O. Tapia and O. Goscinski, Mol. Phys. (1975) 29, 1653. O. Tapia in *Quantum Theory of Chemical Reactions*, eds. A. Pullman, L. Salem, A. Veillard, Reidel, Dordrecht (1981) Vol 2.
[30] M.M. Karelson, A. R. Katritzky and M. C. Zerner, Intern. J. Quantum Chem. (1986), Symp 20, 521.
[31] M. M. Karelson and M. C. Zerner, J. Phys. Chem. (1992) 96, 6449.

[32] This is an easy term to derive by considering a Born - Haber cycle, and charging a sphere of radius ao in the gas phase ΔG(gas) and then in a medium of dielectreic ε, $\Delta G(\varepsilon)$. Then ΔG (ion solvation) = - $\Delta G(\varepsilon)$ + ΔG(cavitation)+ ΔG(gas), where ΔG(cavitation), the energy to create a cavity in the dielectric medium, is small compared to the other values. This leads to ΔG(ion solvation) = $1/(2\varepsilon$ ao) + 0 - 1/ (2 ao) = -1/2 (1-1/ε) ao^{-1}. See reference 36 below.

[33] J. G. Kirkwood, J. Chem. Phys. (1934) 2, 351 ; J. G. Kirkwood and F. H. Westheimer, J. Chem. Phys. (1938) 6, 506.

[34] L. Onsager, J. Am. Chem. Soc. (1936) 58, 1486.

[35] E. Lippert, Ber. Bunsen. Ges. Phys. Chem. (1957) 61, 562

[36] See C. J. F. Böttcher, *Theory of Electric Polarization,* 2nd ed., Elsevier, Amsterdam (1973) Vol.1 and 2 for a good classical treatment of this subject.

[37] E. G. McRe, J. Phys. Chem. (1957) 61, 562.

[38] P. O. J. Scgerer and S. F. Fischer, Chem. Phys. (1989) 131, 115

[39] R. G. Alder, W.W. Parson, Z. T. Cu and A. Warshel, submitted for publication.

[40] M. Marchi, J. N. Gehlen, D. Chandler and M. Newton, J. Am. Chem. Soc. (1993) 115, 4178.

[41] J. N. Gehlen, M. Marchi and D. Chandler, in press.

[42] M. Cory and M. C. Zerner, submitted for publication

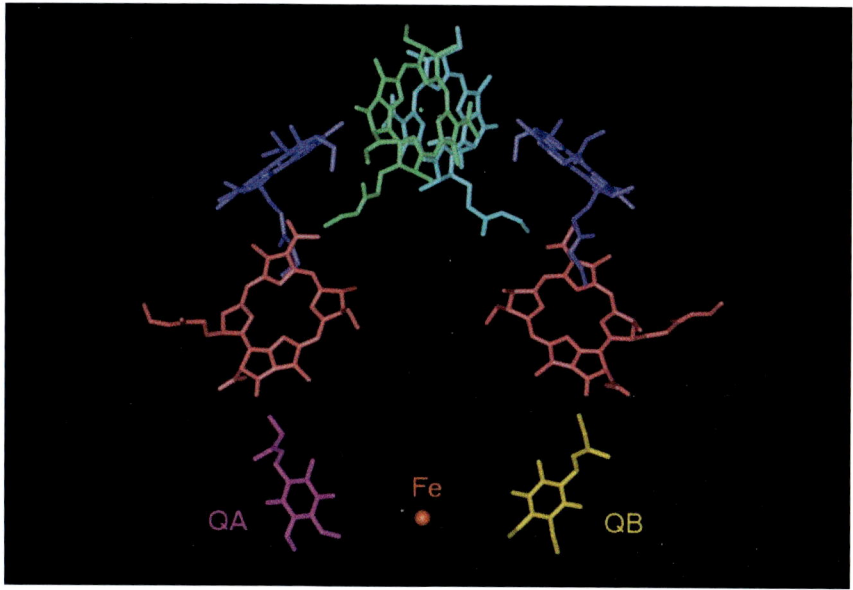

Figure 2. Arrangement of the electron transfer cofactors in the photosynthetic reaction center protein from the bacterium *Rhodobacter sphaeroides*. The figure shows the special pair of bacteriochlorophylls (top, in green and light blue), two accessory bacteriochlorophyll molecules (dark blue), two bacteriopheophytins (red), the primary quinone (Q_A), the secondary quinone (Q_B), and the non-heme iron.

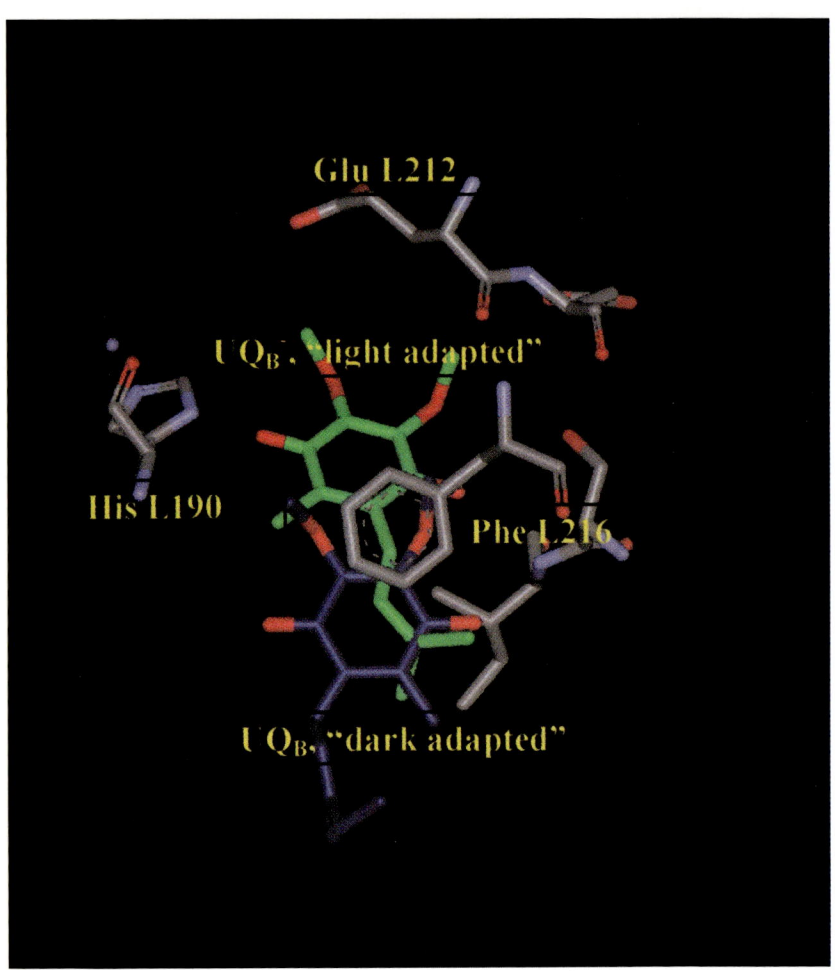

Figure 6. Amino acid side chains near the secondary ubiquinone binding site in the light adapted structure (green) of the Rb. sphaeroides RC. One oxygen of ubiquinone is within hydrogen bonding distance of the His L190 side chain and the Glu L212 side chain is oriented further away. Also shown is the binding site of UQ_B in the dark adapted structure (blue) and the Phe L216 side chain (Created from coordinates reported in reference 14. Isoprenyl chains of the ubiquinones have been truncated for clarity).

Chapter 3

Molecular Modeling and Simulation of a Reaction Center Protein

Matteo Ceccarelli[1], Marc Souaille[1,2], and Massimo Marchi[3]

[1]Centre Européen de Calcul Atomique et Moléculaire (CECAM), Ecole Normale Supérieure de Lyon, 46 Allée d'Italie, F–69364 Lyon Cedex 07, France
[2]Physical Chemistry Institute, University of Zürich, Winterthurerstrasse 190, CH–8057 Zürich, Switzerland
[3]Section de Biophysique des Protéines et des Membranes, DBCM, CEA–Saclay, F–91191 Gif-sur-Yvette Cedex, France

Abstract

In this paper we give an account of our ongoing effort to understand bacterial photosynthesis at the atomic level. First, we describe earlier simulations which investigate the nuclear motion coupled to the primary donor excitation in bacterial reaction centers (RC). Then, we discuss the molecular modeling of the chromophores of the RC of *rhodobacter sphaeroides*. Finally, we report on our latest molecular dynamics simulation results concerning a RC in a detergent micelle.

The bacterial photosynthetic reaction center (RC)[1, 2] is a membrane protein composed of chromophores (bacteriochlorophylls, bacteriopheophytins and quinones) and three protein subunits named L, M and H. While proteins L and M form two branches of the RC (almost the mirror images of each other) and provide the necessary scaffolding to hold in place bacteriochlorophylls and bacteriopheophytins, the H subunit is in contact with the bacterial cytoplasm and binds the quinones in its interior. In the region of the RC near the pery-plasm is located a bacteriochlorophylls dimer, the so-called special pair (P). This chromophore is at the junction point between the L and M branches and is involved directly in the first photosynthetic electron transfer.

Photosynthesis in purple bacteria commences with the excitation of P, the energy for such excitation being transferred directly from surrounding light harvesting proteins. The excited state P* then decays in a charge transfer state, $P^+ H_L^-$, where an electron has been exchanged with the bacteriopheophytins on the L branch (H_L), 17 Å away. The electron transfer (ET), $P^* H_L \to P^+ H_L^-$, occurs with a quantum yield of 1 and is called primary charge separation. Experimentally no electron is detected arriving to the bacteriopheophytins on the M branch, H_M. Subsequently, the electron on H_L is first, transferred to the quinone (whose chemical identity changes with the type of RC) on the L side, or Q_A, and second, passed to the quinone on the M side, or Q_B.

The primary ET in RC presents many aspects which are very challenging to understand. Despite the many efforts, as yet unresolved by either experiment or theory are the following features: (a) the remarkably fast rate ($\tau^{-1} \sim 3$ ps) of this electron exchange between two chromophores separated by 17 Å; (b) the fact that the M branch acts as spectator in the reaction, despite structural similarities and quasi-C_2 symmetry between the L and M subunits; (c) the role of the accessory bacteriochlorophyll in Van der Waals contact with P and H_L in the mechanism of ET.

Since the first crystal structures of bacterial photosynthetic RC have been resolved[1], there have been a few investigations of the primary ET in the photosynthetic reaction center by MD simulations[3-6]. All of these investigations are based on a reduced quantum mechanical model whose relevant parameters are derived by running classical MD. This is the so-called spin–boson model[7]. In the last 10 years or so, powerful techniques, derived mostly from this simple model, have been developed to handle linear and non–linear spectroscopy[8]. This entails using the so–called spectral density formalism which relates spectroscopic properties to the fluctuation in the energy gap (i.e., energy difference) function between the ground and the excited states. ET theories derived from the Marcus approach[9] and based on the same spectral density formalism have also been proposed in the past[7]. In that case the energy gap is the energy difference between the neutral and the charge transfer states.

This paper is an account of our continuing effort in the investigation of the primary charge separation in bacterial RC. In the next sections we will describe in some detail our advances in the modeling and understanding of the RC photosynthetic proteins.

Nuclear Coupling in the $P \to P^*$ Transition

Due to limitations in computational power, the first simulations of RC proteins either ignored[4, 5] or used mean field approximations[3, 6] of the environment surrounding these proteins. In both cases, however, only the atoms closer to the cofactors were actually simulated while the remaining atoms were kept fixed or harmonically restrained. In Ref. [10] we have improved on this model by simulating a RC of *rhodobacter (rb.) sphaeroides* in water. In practice, the simulation box of dimensions $a = 65.63$ Å , $b = 58.0$ Å and $c = 65.0$ Å was composed of a RC protein having its quasi-symmetry axis parallel to the a-direction and of 4104 water molecules which filled in the voids. All the simulation runs in Ref. [10] were carried out by using our in-house molecular dynamics program ORAC[11, 12]. In this earlier work, a spherical cutoff of 9 Å was applied to the non-bonded interactions thus neglecting long range electrostatic effects. The potential parameters and the molecular topology of the system is fully described in Ref. [10]. Simulations of about 100 ps each were carried out at 300 K and 50 K. The latter calculation was done to compare with time resolved data obtained at low temperatures ($\simeq 10$ K)[13].

Our investigation on a hydrated reaction center was focused on the the dynamics of the P → P*, transition which occurs before primary charge separation. In the past, time resolved stimulated emission studies in the P → P* region have been carried out on different RC[13, 14]. A common feature obtained in all these studies is that low frequency (less than 200 cm^{-1}) nuclear modes are coupled to this photoexcitation. Although it has been shown that the environment surrounding the special pair is responsible for the coherent oscillations found in the spectra[13], no indication on the atomistic origin of these vibrations has been found.

The principal aim of our study was to gain insights in the coupling mechanism between nuclear modes and electronic transition. To achieve this goal, we computed the time–resolved stimulated emission spectra in different regions of the P* → P transition. This involved carrying out molecular dynamic simulations of the hydrated bacterial reaction center of *rb. sphaeroides* in the ground and excited states of the special pair.

Modeling Photoexcitation

The spin–boson model is well suited to handle theoretically photoexcitation of a pigment molecule in contact with a dielectric from a ground to an excited state. In this model, the two electronic states of the chromophores are represented by a two–level system which is linearly coupled to a fluctuating dielectric described by harmonic fields. The interaction between dielectric and the photoexcitation is responsible for the broadening of the pigment electronic bands in solution or within a protein. It is worth to point out that even for complex systems the restriction on the harmonicity of the field is not so constraining as it might seem. Indeed, the binding requirement here is that only the nuclear modes coupled to photoexcitation be harmonic[5].

This is in general compatible with very anharmonic systems such as that discussed here.

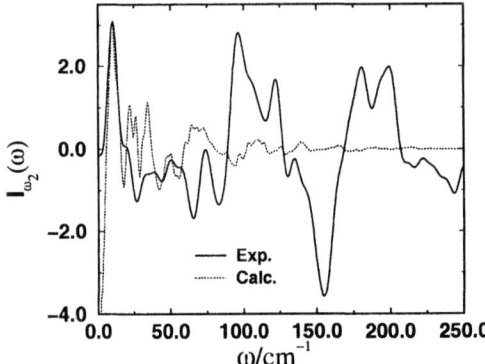

Figure 1: Comparison between the experimental and calculated reduced spectral densities. The two spectra are normalized with respect to the intensities of their lowest frequency peak.

In time resolved stimulated emissions experiment[13] the sample is at first pumped to the excited state by a test beam of light of a given frequency ω_1. Then, at variable time delays, a white light continuum beam is shone on the photoexcited sample to probe its absorption spectrum. In the end, the experiment measures the time resolved transmission induced by the photoexciting beam of frequency ω_1 on the absorption at a given frequency ω_2. This transmission is then converted to absorption changes which in turn are proportional to the stimulated emission coefficient $S(\omega_1, \omega_2, t)$. Using a displacedharmonic oscillator model, Mukamel[8] has shown that this emission coefficient can be readily computed from the knowledge of the energy gap dynamics. Thus, the major goal of our molecular dynamics investigation was to compute the energy gap for the P* → P transition as a function of time. Although empirical charge distribution was used in our study, *ab initio* data can, in principle, be obtained to model the charge distribution in the P and P* states. As for the pseudopotential model for the interaction between the special pair and the environment (proteins and water), we used a purely electrostatic model.

Results

After 130 ps of molecular dynamics equilibration and 100 ps of additional run time, the deviation from the X–ray structure (X_{rms}) of the three subunits increased to an average of 3.1 Å for the α carbons. The subunit H had the largest deviations, while the membrane-spanning helices had a relatively low value of X_{rms}. Of the 4101 water molecules solvating the RC, only 304 were trapped near the protein for more than 90 % of the time. They were found more often in the hydrophilic region of the subunits than near the membrane–spanning helices of the L and M subunits.

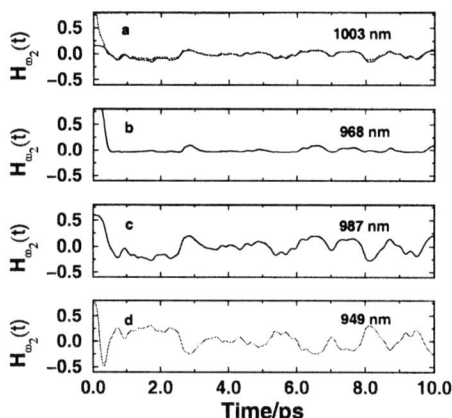

Figure 2: Calculated time resolved stimulated emission intensity. From top to bottom the four panels present results for probing frequency ω_2 equal to respectively (a) 1003 , (b) 968, (c) 987 and (d) 949 nm. For comparison, the autocorrelation function of the energy gap has been added on the panel (a) (dotted line).

During the 90 ps run at 50 K in the P* state, the energy gap or energy difference between the P excited and ground state were accumulated. The energy gap probability distribution derived from this simulation obeyed, to a good approximation, Gaussian statistics. This result gives credibility to our approach in the calculation of the electronic spectra based on a linear response model.

Our calculation found that the oscillatory part of $S(\omega_1, \omega_2, t)$ at ω_1 (or $H_{\omega_2}(t)$) behaves like the experimental time evolution of the stimulated emission spectra. As shown in From Fig. 2, away from the bottom of the emission peak ($\omega_2 = 968$ nm), $H_{\omega_2}(t)$ shows an oscillatory behavior, with a phase change of π between ω_2=987 and 949 nm. Noticeably, at ω_2=968 nm, the maximum of the emission peak, the aspect of $H_{\omega_2}(t)$ changes showing a decrease in the amplitude of the oscillations as well as the appearance of different oscillating features.

For all wavelength, $H_{\omega_2}(t)$ shows coherent oscillations well beyond the 3 ps time mark. This contrasts with experimental results showing an exponential damping of the time correlation function within this time scale. Such a damping might be due to frictional damping or static disorder on the RCs.

The experimental and calculated reduced spectral densities obtained by Fourier transform of $H_{\omega_2}(t)$ are compared in Fig. 1. Although the calculation reproduces the first experimental peak at 10 cm^{-1}, the activities of the higher frequency modes is underestimated with respect to experiment. This result and the finding in Ref. [10], that the higher frequency peaks (at 67 and 125 cm^{-1}) in the dimer energy gap spectral density are weakened considerably by solvent effects, may suggest the our simple electrostatic model underestimates the dimer-solvent energy gap with respect to the intra dimer counterpart.

Improving the Reaction Center Model

Starting from the earlier simulation[10] described above, we have in the last few years, achieved major improvements on the modeling of the RC proteins and their environment. Indeed, MD simulations of membrane proteins are challenging for different reasons. First, although force fields for proteins have been steadily improving, *ab initio* based potential models for chromophores such as bacteriochlorophylls and quinones have been lacking. It is clear that in studies of ET realistic modeling of the prosthetic groups is essential. Second, because the coupling between electronic states and the solvent (the surrounding protein) is at the first order electrostatic, electrostatic forces must be computed accurately in MD simulations. Because of the high cost this is not done in standard biomolecular simulations and was not done in our earlier simulation. Third, the large size of the system: the RC protein of *rb. sphaeroides* by itself contains more than 15,000 atoms. The size of the system increases considerably if its natural environment, solvent and phospholipid membrane, is included in realistic simulations. Last but not least, the issue of electronic polarizability in modeling RC's will need to be addressed because it is relevant to electron transfer.

Modeling the Chromophores

Deriving a force field for chromophores entails calculations of the electronic structure of these large molecules. In recent times, the density functional theory (DFT)[15] approach has been gaining popularity in the chemical community to investigate on the electronic ground state structure of chemically relevant system. In the past, this technique has been applied by us and others to the study of bacteriochlorophylls molecules[16, 17] and quinones[18]. All our DFT calculations on the chromophores were carried out in the local density plus gradient corrections approximation (LDA+GC)[19], while the Kohn-Sham orbitals were expanded in a basis set of plane waves compatible with the periodic supercell containing the molecule. In this scheme only the valence electrons are included explicitly in the calculation. The effects due to atomic core electrons are taken into account by *ab initio* soft pseudopotentials associated with each atom[20].

In an earlier calculation[16] on a crystal of methyl bacteriopheophorbide (MeBPheo) *a* we have shown the viability of the DFT technique towards the computation of the electronic structure and optimization of molecules important for photosynthesis. In a more recent investigation[17] we have carried out the first (to our knowledge) high quality DFT *ab initio* calculation of the vibrational structure of a bacteriochlorophyll derivative and assigned the in plane frequencies detectable by resonant Raman spectroscopy. The *ab-initio* calculations were performed on an isolated methyl bacteriochlorophyll *a* molecule. Our major result was the determination of a complete set of eigenvalues and eigenvectors which unambiguously identified the fundamental modes. The computed frequencies are in very good agreement with the available experimental data and show a small root mean square deviation (less than 20 cm^{-1}) from experimental modes. In addition, our calculation was able to pin point a strong symmetric behavior of many in-plane vibration similar to results obtained for porphyrines. This is in contrast with results of earlier semiempirical calculations[21], which predicted more localized modes in (bacterio)pheophorbide-type macrocycles.

Based on DFT calculations on chlorophylls and, additionally, on ubiquinone and the RC main detergent, lauryl dimethylamine oxide or LDAO, we have then developed a force field for their classical modelization. Our approach to this undertaking was straightforward. We initially use the DFT optimized structures and the vibrational analysis to determine the bonded part of the potential parameters described by the AMBER potential function. Then, atomic *ab initio* partial charges on the chromophore are used to account for electrostatic effects. At a later stage, experimental data from X-ray crystallography are used to check the structural properties of the molecule in the condensed state and to refine the intermolecular Lennard-Jones parameters.

A good starting point for parameter refinement of large molecules are values optimized on small fragments. In the case of bacteriochlorophylls, we started modeling their chemical constituents such as pyrrol, methyl acetate and methylic groups, for which DFT calculations were also made. Sub-

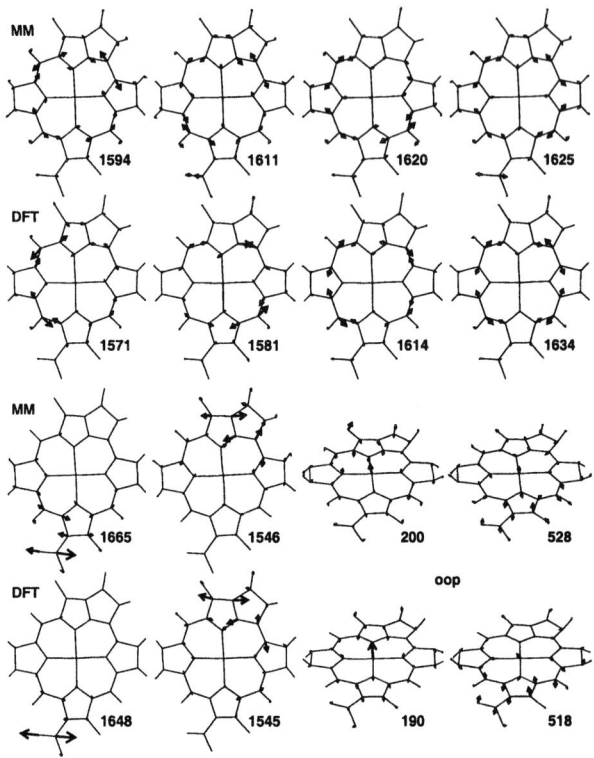

Figure 3: Methyl bacteriochlorophyll *a* normal modes. Frequencies are in cm^{-1}

sequently, these parameters were transferred to the larger molecules and refined as much as possible to reproduce the bacteriochlorophylls DFT results. The existing AMBER parameters were used, whenever possible, as a starting point for potential of the fragments. Given the relative smaller size of the quinone and the hydrophilic head of LDAO[1], the force field refinement was done in this case directly on the molecules and not the fragments.

In all our molecular modeling, special care was taken not only to reproduce the DFT frequencies, but also the eigenvectors. Indeed, a great deal of spectroscopic experiments on photosynthetic RC involve directly or indirectly the nuclear dynamics of the chromophores. In Fig. 3 we present

---

[1] DFT optimization and vibrational analysis were performed on models of ubiquinone and LDAO obtained by cutting and saturating their tails after the first 3 carbon atoms.

some typical in–plane and out-of–plane normal modes computed from DFT and molecular mechanics (MM). Its visual inspection shows the good agreement between DFT and MM eigenvectors.

Methods for Molecular Dynamics

Due to the ever increasing power of computers, simulations of solvated proteins for lengths up to hundreds of ps is nowadays becoming routine on desktop workstations. However the simulation of membrane proteins stabilized in membrane or detergent environments is more challenging given the complexity and the large size of such systems. In addition, the dielectric properties of the aqueous solvent are of crucial importance to electron and proton transfer reactions. Consequently, any theoretical investigation of such phenomena must deal with the problem of how to simulate large systems and, at the same time, correctly handle long-range electrostatic interactions.

In the past few years, we have developed new and fast multiple timestep MD algorithms[11, 22, 23] which allow simulations of very large systems and include an accurate representation of the Coulombic interactions for infinite systems. The electrostatic series can be computed in principle *exactly* using the Ewald re-summation technique[24]. This method in its standard implementation, is extremely CPU demanding and scales like N^2, with N being the number of charges with the unfortunate consequence that even moderately large size simulations of inhomogeneous biological systems are not within its reach. Notwithstanding, the rigorous Ewald method, which suffers of none of the inconvenients experienced by the reaction field approach, has very recently regained the spotlight with the particle mesh Ewald (PME) technique[25]. PME, based on older ideas of Hockney's[26], is essentially an interpolation technique which involves charge smearing onto a regular grid and evaluation of the reciprocal lattice energy sums via fast Fourier Transform (FFT). The performance of this technique, both in accuracy and efficiency, is very good. Most importantly, the computational cost scales like $NlogN$ for large N, but it is essentially *linear* for practical application up to 20,000 particles.

The combination of the multiple time step algorithm and PME[27] makes the simulation of large size biomolecular systems such as membrane proteins extremely efficient and affordable even for long time spans. Furthermore, it does not involve any uncontrolled approximation and is entirely consistent with periodic boundary conditions.

Our most recent r-RESPA algorithm in combination with constraints and PME achieves substantial computer savings (more than 10 times) with respect to standard Ewald techniques. In spite of this excellent performance, for a 40,000 atom system 1 ns of simulation takes more than 12 days on a fast Compaq EV5 station. Further improvements can be achieved running simulation on parallel machines. Indeed, a parallelized version of ORAC has been recently developed and each ns of simulation for the same systems now takes 3 days on 8 processors of a Compaq ES232 cluster.

Simulation of a RC in a Detergent Micelle

To elucidate at the atomic level the asymmetry and the mechanism of the primary charge separation it is advantageous to construct a reliable model of the RC protein and of its environment. Given the latest technical

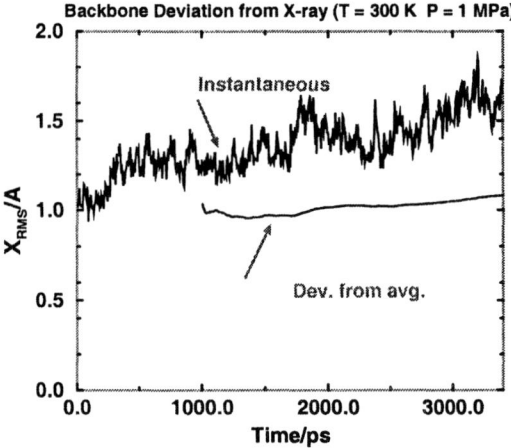

Figure 4: Deviation from the crystallographic structure of the RC protein during 3.4 ns of simulation. The upper curve is the instantaneous deviation, while the lower curve is the deviation of the averaged structure.

developments in our computational tools[23], we have recently been able to simulate the RC of *rhodobacter sphaeroides* inglobed in a model micelle of detergent lauryl-dimethyl-amine-oxide (LDAO) and hydrated by water for more than 3.4 ns. This hydrated micelle was constructed as follows. First, the isolated RC protein[28] including 9 LDAO molecules and 160 crystallization waters were relaxed at 20 K for about 5 ps to eliminate any accidental stress due to the force field. To this structure we then added 150 LDAO molecules initially disposed around the protein hydrophobic helices in an all–trans conformation and with the polar head pointing away from the protein assembly. From this conformation we ran equilibrations of the

LDAO only (i.e. keeping fixed the protein coordinates) for 30 and 230 ps at 250 K and 400 K, respectively. After this step the added LDAO had relaxed to form an aggregate around the RC in which the aliphatic chains were in contact with the transmembrane helices and the polar heads pointed towards the exterior.

Finally, the system was placed in an orthorhombic box of dimension of 76x84x80 Å and hydrated by 6323 additional water molecules to fill the cell voids. A simulation run on our final system, totaling 40,327 atoms, was then started at constant temperature and pressure at T=300 K and P=0.1 MPa.

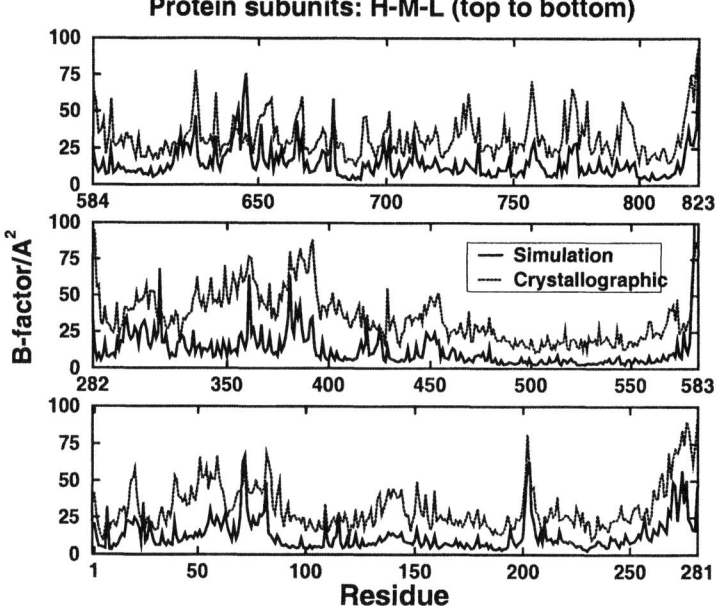

Figure 5: Comparison between calculated and crystallographic Debey–Waller factors.

To achieve full relaxation the simulation box was entirely flexible for the first 300 ps and evolved to a hexagonal-like structure. For the remaining runs only isotropic changes of the box were allowed. With the simulation algorithms in the NPT ensemble described in Ref. [29] and on 32 processors of a T3E parallel computer, 3.4 ns of simulation required 21 days of simulation.

In this report we only provide a brief account of some preliminary results on the structure of the simulated RC. The full account of our results will be given in future publications.

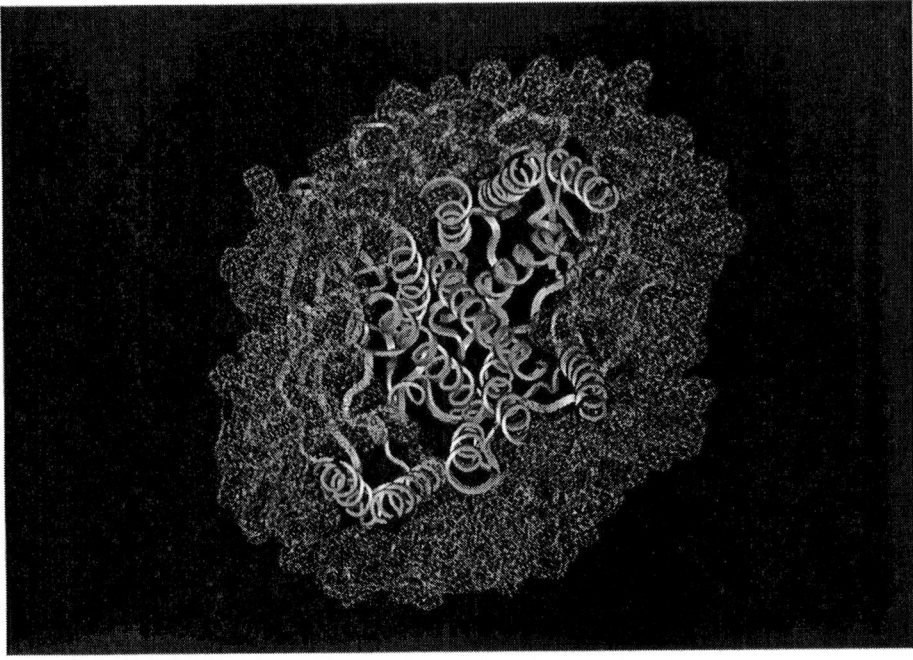

Figure 6: Instantaneous view of the RC of *rb. sphaeroides* during the simulation (see text for details).

Fig. 4 shows the overall deviation (or X_{rms}) of the RC backbone atoms from their X-ray coordinates. We first notice that with respect to our previous, and much shorter, simulations the X_{rms} is twice as small. Moreover, the RC protein shows with time an overall drift towards higher X_{rms}. With additional investigation, we find that this is due essentially to relative motion of subunit H with respect to L and M.

In addition to a low deviation from the crystallographic structure, the experimental Debey-Waller factors are also well reproduced within a constant offset, probably due to disorder (see Fig. 5).

In Fig. 6 we show a pictorial view of the RC protein surrounded by its detergent. The LDAO molecules are represented by their molecular surface and no water molecules are displayed. This picture is evocative of the detergent micellar structure revealed by the low resolution neutron scattering study of the RC crystal[30]. As in that investigation, our detergent has formed a distinct micelle around the protein. Its thickness along the quasi-C_2 symmetry axis is around 25-30 Å, similar to that observed experimentally. We point out that in the simulation the micelle is a rather dynamics structure with a lateral diffusion constant of 5.6×10^{-10} m^2 s^{-1} which is about 30 times larger than that of the RC protein.

Conclusion and Perspectives

Thanks to our ongoing efforts in molecular modeling of RC proteins, we have recently devised a reliable force field and a faithful model for photosynthetic RC proteins in detergent environment. Long simulations (3.4 ns) of a RC protein of *rb. sphaeroides* in its detergent environment have also been possible due our recent technical developments in the MD methodology. The analysis of properties related to the primary ET and its asymmetry is currently underway. Our first task is to compute the average electrostatic field and its dynamics on the bacteriochlorophyll chromophores of the L and M branches to investigate the origin of the ET asymmetry. In this calculation we include both contributions from the fixed charges of the RC and from induction computed from a distributed polarizable model[31]. Investigation of effects from mutations is also underway.

References

[1] Deisenhofer, J.; Epp, O.; Mike, K.; Huber, R.; Michel, H. *J. Mol. Biol.* **1984**, *180*, 385.
[2] Deisenhofer, J.; Michel, H. *Science* **1989**, *245*, 1463.
[3] Creighton, S.; Hwang, J. K.; Warshel, A.; Parson, W. W.; Norris, J. *Biochemistry* **1988**, *27*, 774.
[4] Treutlein, H.; Schulten, K.; Brunger, A.; Karplus, M.; Deisenhofer, J.; Michel, H. *Proc. Natl. Acad. Sci. USA* **1992**, *89*, 75.
[5] Marchi, M.; Gehlen, J. N.; Chandler, D.; Newton, M. *J. Am. Chem. Soc.* **1993**, *115*, 4178.
[6] Parson, W. W.; Chu, Z. T.; Warshel, A. *Photosynthesis Research* **1998**, *55*, 147.
[7] Chandler, D. In *Liquids, Freezing and the glass transition*, Leveque, D.; Hansen, J. P., edts. North Holland, 1991, Page 193.
[8] Mukamel, S. *Principles of Nonlinear Optical Spectroscopy.* Oxford University Press, **1995**.
[9] Marcus, R. A.; Sutin, N. *Biochim. Biophys. Acta* **1985**, *811*, 256.
[10] Souaille, M.; Marchi, M. *J. Am. Chem. Soc.* **1996**, *119*, 3948.
[11] Procacci, P.; Darden, T.; Paci, E.; Marchi, M. *J. Comp. Chem.* **1997**, *18*, 1848.
[12] Procacci, P.; Marchi, M. In *Advances in the Computer Simulations of Liquid Crystals*, Zannoni, G.; Pasini, P., edts, NATO ASI School (Kluwer Academic publishers, Dordrecht the Netherlands, 1999).

[13] Vos, M. H.; Rappaport, F.; Lambry, J.-C.; Breton, J.; Martin, J.-L. *Nature* **1993**, *363*, 320.
[14] Vos, M. H.; Jones, M. R.; Hunter, C. N.; Breton, J.; Lambry, J.-C.; Martin, J.-L. *Biochemistry* **1994**, *33*, 6750.
[15] Parr, R. G.; Yang, W. *Density Functional Theory of Atoms and Molecules*. Oxford University Press, Oxford **1989**.
[16] Marchi, M.; Hutter, J.; Parrinello, M. *J. Am. Chem. Soc.* **1996**, *118*, 7847.
[17] Ceccarelli, M.; Lutz, M.; Marchi, M. *J. Am. Chem. Soc.* **2000**, *122*, 3532.
[18] Nonella, M.; Brändli, C. *J. Phys. Chem.* **1996**, *100*, 14549.
[19] We use the interpolation of Perdew, J. P.; Zunger, A. *Phys. Rev. B* **1981**, *23*, 5048 for the electron gas correlation energy. We use the gradient correction of Becke for exchange (Becke, A. D. *J. Chem. Phys.* **1986** *84*, 4524), and of Perdew (Perdew, J. P. *Phys. Rev. B* **1986** *33* 8822) for correlation.
[20] Troullier, N.; Martins, J. L. *Phys. Rev. B* **1991**, *43*, 1993.
[21] Donohoe, R. J.; Frank, H. A.; Bocian, D. F. *Photochem. Photobiol.* **1988**, *48*, 531.
[22] Procacci, P.; Darden, T.; Marchi, M. *J. Phys. Chem.* **1996**, *100*, 10464.
[23] Procacci, P.; Marchi, M. In *Advances in the Computer Simulations of Liquid Crystals,* Zannoni, G.; Pasini, P., edts, NATO ASI School (Kluwer Academic publishers, Dodrecht the Netherlands, 1998).
[24] de Leeuw, S. W.; Perram, J. W.; Smith, E. R. *Proc. R. Soc. Lond.* **1980**, *A 373*, 27.
[25] Darden, T.; York, D.; Pedersen, L. *J. Chem. Phys.* **1993**, *98*, 10089.
[26] Hockney, R. W. *Computer Simulation Using Particles*. McGraw-Hill, New York, **1989**.
[27] Procacci, P.; Marchi, M. *J. Chem. Phys.* **1996**, *104*, 3003–3012.
[28] Ermler, U.; Fritzsch, G.; Buchanan, S. K.; Michel, H. *Structure* **1994**, *2*, 925.
[29] Marchi, M.; Procacci, P. *J. Chem. Phys.* **1998**, *109*, 5194.
[30] Roth, M.; Lewit-Bentley, A.; Michel, H.; Deisenhofer, J.; Huber, R.; Oesterhelt, D. *Nature* **1989**, *340*, 659.
[31] Thole, B. T. *Chem. Phys.* **1981**, *59*, 341.

Chapter 4

Simulating Thermochemistry of *p*-Benzo-quinone Reduction and Binding of Ubiquinone in the Photosynthetic Reaction Center

Ralph A. Wheeler

Department of Chemistry and Biochemistry, University of Oklahoma, 620 Parrington Oval, Room 208, Norman, OK 73019 (telephone: 405-325-3502, e-mail: rawheeler@chemdept.chem.ou.edu)

This contribution reviews quantum chemical and molecular dynamics simulations of *p*-benzoquinones, including the ubiquinones, plastoquinones, and menaquinones involved in photosynthesis. B3LYP-type hybrid Hartree-Fock/density functional methods give accurate structures, vibrational frequencies, and electron affinities for the *p*-benzoquinones. Force fields derived from these calculations serve as input to molecular dynamics (MD) simulations to calculate one-electron reduction potentials with an average absolute difference of 0.06 eV from experiment for a group of eight *p*-benzoquinones. Conformational changes for a key methoxy group upon reducing ubiquinone provides a way for proteins to modulate electron transfer *rates* as well as thermodynamics by tuning the methoxy torsional angles of ubiquinones. MD simulations of ubiquinone-10 binding in the photosynthetic reaction center of *Rhodobacter sphaeroides* is described and a protein conformational gate controlling ubiquinone binding, migration, and perhaps electron transfer, is identified.

© 2004 American Chemical Society

Introduction

Free energy used by most biological systems originates from solar energy stored by photosynthesis in green plants, algae, or photosynthetic bacteria. The fundamental process of photosynthesis(1) may be represented by one deceptively simple chemical reaction:

$$nH_2O + nCO_2 + light \rightarrow (CH_2O)_n + nO_2 \qquad (1)$$

Hidden within this equation for using light to synthesize carbohydrates from carbon dioxide and water are myriad intermediate steps, mediated by multiple proteins and small molecules.(2) The primary photochemical events implicit in equation (1) involve energy conversion and storage by separating charge across a membrane. In plants, algae, and photosynthetic bacteria, charge separation is accomplished in different protein-pigment complexes that couple two successive one-electron reductions of quinone electron acceptors (see Figure 1) to proton transfer, thereby establishing a proton gradient across the membrane. The proton gradient drives subsequent energy storage by adenosine triphosphate (ATP) synthesis, and the ATP is later used in carbon-fixation reactions to make carbohydrates.

In green plants photosynthesis occurs within the thylakoid membranes of organelles called chloroplasts. First, light excites chlorophyll molecules housed in light-harvesting "antennas" and the energy is passed to a "reaction center". Next, the reaction center uses the electronic excitation energy to separate charge across the thylakoid membrane. Plants and algae accomplish charge separation through the interplay of two photosystems—photosystem I and photosystem II. Photosystem II reduces a quinone, Q, and oxidizes water to O_2:

$$2Q + 2H_2O + light \rightarrow O_2 + 2QH_2 \qquad (2)$$

QH_2 then feeds its electrons into photosystem I via a cytochrome b_6f complex and simultaneously loses two protons. Two high-energy electrons are subsequently shuttled to photosystem I by plastocyanin and the electrons are then passed to NADP reductase by ferredoxin. NADP reductase uses the electrons to reduce $NADP^+$ to NADPH. The proton gradient formed in the process is also used to synthesize ATP from ADP. Ultimately, energy stored as NADPH and ATP is used in carbon fixation reactions to make carbohydrates.

Although cyanobacteria also use two photosystems, most photosynthetic bacteria have only a single photosynthetic reaction center

(RC).(3) Purple bacteria, for example, carry out cyclic electron transport, generate a proton gradient, and synthesize ATP. Bacterial reaction centers from *Rhodobacter (Rb.) sphaeroides* and *Rhodopseudomonas viridis* contain similar electron transfer cofactors as the RC of photosystem II: a "special pair" of bacteriochlorophylls, two bacteriopheophytins, two bound quinones, and an iron bound to four histidine nitrogens and two oxygens of a glutamate. The protein composition of the bacterial RC consists of three polypeptides, denoted L (light), M (medium), and H (heavy). Because bacterial photosynthetic RCs show many structural and functional homologies with reaction centers from green plants,(4-7) bacterial RCs are used extensively as models for plant RCs.

Within the bacteria RCs, photoexcitation of the RC drives single electron transfer from the special pair of bacteriochlorophylls to the primary electron acceptor, one of two bacteriopheophytin molecules. The reduced bacteriopheophytin transfers the electron on to a primary ubiquinone-10 acceptor (see Figure 1), UQ_A, to form the semiquinone anion, UQ_A^-. Then, UQ_A^- transfers the electron to a secondary ubiquinone-10, UQ_B, to form UQ_B^-. After the initial charge separation, the special pair is re-reduced, UQ_B^- accepts a

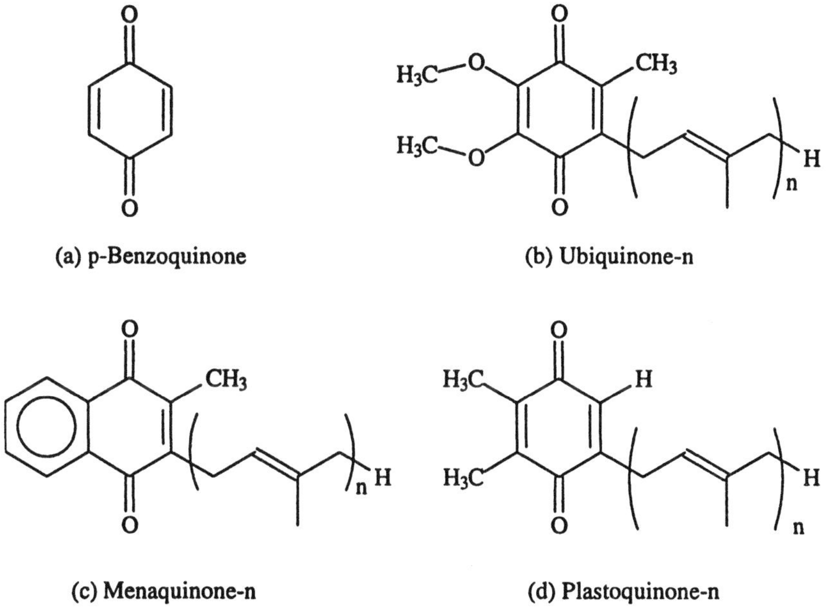

(a) p-Benzoquinone

(b) Ubiquinone-n

(c) Menaquinone-n

(d) Plastoquinone-n

Figure 1. The parent p-benzoquinone molecule and other p-benzoquinones important in photosynthesis.

proton, and in a similar reduction cycle, UQ_BH accepts a second electron and proton to form its hydroquinone form, UQ_BH_2. The hydroquinone is loosely bound to the protein and is in equilibrium with a pool of ubiquinones. A ubiquinone from the pool replaces UQ_BH_2 and regenerates the original reaction center.

X-ray diffraction structures of the photosynthetic RCs from *Rb. sphaeroides*(8-10) and *Rhodopseudomonas viridis*(11) support the inferred similarities between the protein folding and quinone binding sites in the RC proteins of purple bacteria and plant photosystem II. Structural features of the bacterial RCs are reviewed extensively in the literature,(12,13) so we focus on the UQ_B binding sites most important for work reviewed here. Different X-ray diffraction structures show UQ_B distributed over a range of binding sites spanning approximately 5Å.(13) The authors of one study(14) (see Plate 2) of the *Rb. sphaeroides* RC reported a "dark adapted" structure containing UQ_B and a "light adapted" structure reported to contain UQ_B^-. Plate 2 shows selected amino acid side chains from their X-ray diffraction structure of the UQ_B^- binding site. The binding site of UQ_B is also shown, obtained by superimposing the amino acid main chain atoms of the two structures. Based on the different binding sites observed for UQ_B and UQ_B^-, the authors proposed that the observed UQ_B binding site (the "dark adapted" site) is electron transfer inactive, and UQ_B migrates to an electron transfer active site (occupied by UQ_B^- in their "light adapted" structure), before electron transfer occurs.

Clearly, quinone reduction is a vital step in the reaction sequence of photosynthetic energy storage. To understand ubiquinone reduction in the photosynthetic reaction center of *Rb. sphaeroides*, we undertook a systematic study of quinone properties in isolation, in water, and in the RC protein. Our ultimate goals include determining the UQ_B and UQ_B^- binding sites and how the RC protein modulates UQ_B reduction. This contribution provides a progress report. It begins by reviewing work designed to test quantum chemical methods for reproducing structures, vibrational frequencies, and electron affinities for *p*-benzoquinones and their semiquinone anions. The structures, force constants, and atomic charges derived from the quantum chemical calculations were adopted as force field parameters in molecular dynamics (MD) simulations to calculate hydration free energy differences between the *p*-benzoquinones and their semiquinone anions, as described in the subsequent section. This section also describes combining calculated hydration free energies and electron affinities to estimate one-electron reduction potentials for *p*-benzoquinones in water. Finally we conclude by discussing simulations of binding sites for the secondary ubiquinone-10 in the RC protein. Since a detailed review of calculated structures, vibrations, and spin properties of quinones and the quinoidal radicals generated during the course of photosynthetic charge separation has appeared elsewhere,(15) the next section provides only a brief review of calculated properties of p-benzoquinones.

Properties of p-Benzoquinones and their Semiquinone Anions

In 1995-96, Scott Boesch provided the first tests of various density functional-based quantum chemical methods for calculating the structures, vibrations, and spin properties of p-benzoquinones and their semiquinone radical anions.(16-18) Although several DF-based methods give comparably accurate structures and harmonic vibrational frequencies for p-benzoquinone, p-chloranil, and p-fluoranil, the hybrid HF/DF methods were selected to give a range of properties in particularly good agreement with experiment. For example, Figure 2 shows a comparison of bond distances determined by electron diffraction(19) and B3LYP calculations for p-benzoquinone, the parent molecule of the biologically relevant quinones shown in Figure 1. Figure 2 shows that calculated bond lengths are all within three standard deviations of experiment, the usual standard applied to decide that two experimentally determined structures are identical. The C-C bond distance is within 0.006 Å of the experimental value, the CO bond distance is identical with the experimentally determined distance, and the calculated C=C distance differs from experiment by only 0.001 Å. Subsequent work implies that the B3LYP method gives only slightly less accurate structural parameters for quinones related to the parent p-benzoquinone, including ubiquinones (UQ-n),(18,20) plastoquinones (PQ-n),(21,22) and menaquinones (MQ-n).(23-25)

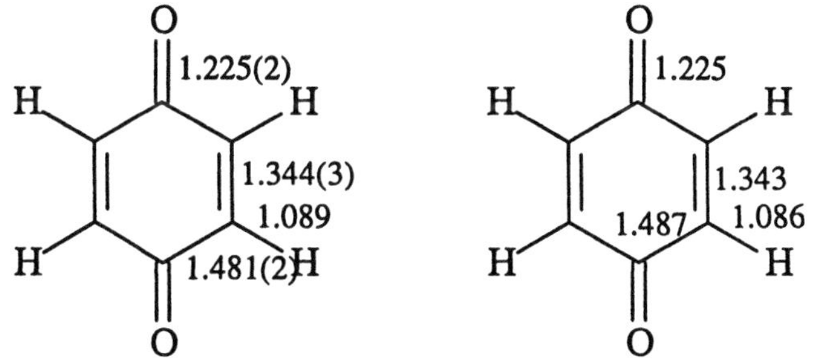

Figure 2. Bond distances for p-benzoquinone (a) determined by electron diffraction(19) and (b) calculated by using the B3LYP/6-31G(d) method.(16)

One-electron reduction of *p*-benzoquinone to give the *p*-benzosemiquinone radical anion gives bond length changes whose direction may be predicted from the LUMO of *p*-benzoquinone.(17,18) Since the LUMO is antibonding along the CO and C=C bonds, CO and C=C bonds are predicted to increase in distance upon occupying the orbital with one (or two) electrons, whereas the C-C single bond distances are predicted to decrease. Thus, frontier orbital arguments agree with the valence bond picture in predicting that *p*-benzosemiquinone becomes more benzenoid upon one-electron reduction (see Figure 3), a prediction confirmed by calculated bond distances. In the *p*-benzosemiquinone anion, CO distances increase substantially, to 1.263 Å, and the C=C bonds distances also increase, to 1.369 Å. As predicted, the CC single bonds of *p*-benzoquinone contract, to 1.449 Å, in *p*-benzosemiquinone radical anion. To a first approximation, the added electron enters the pi system of *p*-benzoquinone and should have minimal effect on CH bond distances, and indeed CH distance increase by only 0.004 Å.

To test the ability of density functional-based methods to provide intramolecular force field parameters needed for MD simulations, vibrational frequencies were also calculated and compared with experiment.(17,18) Table I shows that the B3LYP method gives harmonic vibrational frequencies for *p*-benzoquinone that differ from experimentally determined values(26-28) by an average absolute magnitude of 40 cm^{-1} (3.9%, corresponding to a multiplicative scaling factor of 0.961, in agreement with more extensive studies of many different molecules(29,30)). The calculated, scaled frequency for the antisymmetric, B_{1u} CO stretching mode is 1689 cm^{-1}, compared with the experimental value of 1666 cm^{-1}. The symmetric, A_g CO stretching mode, calculated to appear at 1688 cm^{-1} is similarly close to the experimentally measured frequency of 1663 cm^{-1}. Likewise, the symmetric and antisymmetric C=C stretching modes (of A_g and B_{2u} symmetry, respectively) are calculated to

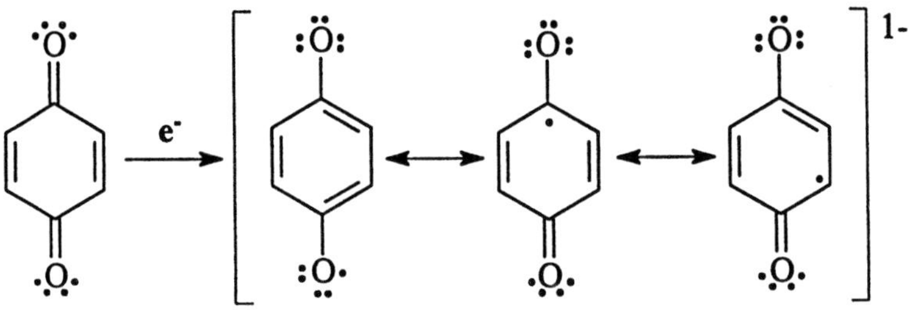

Figure 3. p-Benzoquinone and resonance structures of the p-benzosemiquinone radical anion, demonstrating the more benzenoid character of the anion.

Table I. Approximate mode descriptions and experimentally measured(26-28) and scaled, calculated(16,18) vibrational frequencies(cm^{-1}) of *p*-benzoquinone (PBQ) and *p*-benzosemiquinone radical anion (PBSQ). Calculations were done using the B3LYP/6-31G(d) method.

Sym.	Assignment	PBQ Expt.	PBQ Calc.	PBSQ Expt.	PBSQ Calc.
a_g	C-H stretch	3058	3099		3058
b_{2u}	C-H stretch	3062	3096		3052
b_{3g}	C-H stretch	3057	3080		3032
b_{1u}	C-H stretch	3062	3080		3031
b_{1u}	C=O stretch	1666	1689		1535
a_g	C=O stretch	1663	1688	1435	1488
a_g	C=C stretch	1657	1629	1620	1643
b_{2u}	C=C stretch	1592	1597		1475
b_{3g}	C-C stretch, C-H bend	1388	1353		1416
b_{1u}	C-H bend	1354	1343		1333
b_{2u}	C-C stretch	1299	1277		1200
b_{3g}	C-H bend	1230	1193		1215
a_g	C-H bend	1160	1131	1161	1113
b_{2u}	C-H bend	1066	1049		1038
b_{2g}	C-H wag	1018	985		947
a_u	C-H wag	989	974		942
b_{1u}	C-C=C bend	944	952		970
b_{3u}	C-H wag	882	907		860
a_g	C-C stretch	774	778		833
b_{2g}	Ring chair bend	800	799		801
b_{1u}	C-C stretch	728	758		796
b_{1g}	C-H wag	766	761		774
b_{3g}	C=O bend, C-C stretch	601	603		634
b_{3u}	Ring boat bend	505	515		520
a_g	C-C-C bend	447	456	481	470
b_{3g}	C=O bend	459	456		467
b_{2u}	C=O bend	409	413		392
a_u	C=C-C bend	330	338		394
b_{2g}	C=O chair bend	249	241		328
b_{3u}	C=O boat bend	89	101		139

appear at 1629 cm^{-1} and 1599 cm^{-1}, respectively, compared to their experimentally measured frequencies of 1657 cm^{-1} and 1592 cm^{-1}. So the B3LYP method not only reproduces the relative ordering of these four modes that appear very close together in frequency, but also gives the actual vibrational frequencies with better accuracy than uniformly scaled Hartree-Fock or MP2 frequencies. Disagreements between calculations and experiment for the analogous modes of tetrachloro- and tetrafluoro-p-benzoquinones,(16) along with subsequent, more detailed studies of p-benzoquinone by Nonella and co-workers,(31,32) warn us however that calculated vibrational frequencies and the relative ordering of these four modes are highly dependent on calculated force constants.

For p-benzosemiquinone radical anion, the B3LYP-derived vibrational modes are shifted to lower frequencies than in the neutral p-benzoquinone (see Table I).(17,18) The antisymmetric, B_{1u} CO stretching modes appears at a calculated, scaled frequency of 1535 cm^{-1} and was not experimentally observed. The symmetric CO stretch of A_g symmetry appears at 1488 cm^{-1}, 53 cm^{-1} above the experimentally observed frequency of 1435 cm^{-1}. The two CC stretching modes of A_g and B_{2u} symmetry appear at calculated frequencies of 1643 cm^{-1} (observed at 1620 cm^{-1}) and 1475 cm^{-1} (not detected experimentally), respectively. Thus agreement between experimentally measured vibrational frequencies and calculated, scaled vibrational frequencies is moderately good for p-benzosemiquione radical anion and calculated frequencies shift in the direction implied by weaker, longer CO and C=C bonds upon reduction of p-benzoquinone to form its semiquinone anion. The excellent agreement between calculated and experimental vibrational frequencies for p-benzoquinones and their semiquinone radical anions—and indeed for a series of substituted analogues(18,20-25) including ubiquinones, plastoquinones, and menaquinones—imply that intramolecular force fields determined from B3LYP should be among the most accurate available from quantum chemical calculations.

Thermodynamics of p-Benzoquinone Reduction

Energy changes involved in the one-electron reduction of p-benzoquinones forms the subject of this section.(33) The top reaction in Figure 4 represents the reduction of a generic quinone Q to its semiquinone anion Q$^-$ in aqueous solution, whereas the other reactions illustrate the conceptual decomposition of the reduction into (a) desolvation of Q, (b) gas-phase reduction, and (c) hydration of the Q$^-$ anion. The reduction free energy in water is equal to the sum of free energies for the indirect route from reactants to products in Figure 4, $\Delta G^{o}_{red}(aq) = \Delta G^{o}_{red}(g) + (\Delta G^{o}_{hyd}(Q^{-}) - \Delta G^{o}_{hyd}(Q))$. Calculating the energy change $\Delta G^{o}_{red}(aq)$ is deceptively complex, because accurate quantum chemical calculations incorporating a sufficient amount of solvent are only now becoming available, whereas molecular dynamics simulations with empirical potentials cannot give accurate estimates of energy changes due to changes in covalent bonding brought about by the reduction. Thus, we conceptually decompose the process into three steps and because free energy is a state function, the free energy difference between the products and reactants in Figure 4 is the same, regardless of the path taken. Thus, we calculate the overall reduction free energy as a sum of the gas phase reduction free energy, estimated by using quantum chemical calculations, and the difference in solvation free energies between p-benzoquinone and its anion. The free energy difference between p-benzoquinone and its anion is calculated by using free energy perturbation theory or thermodynamic integration.(34)

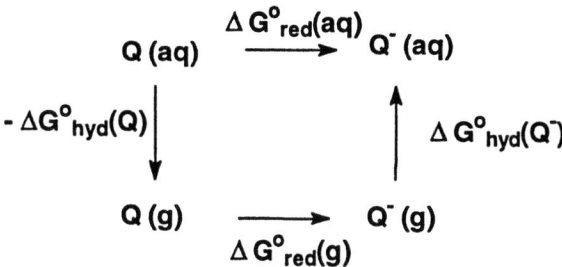

Figure 4. Thermodynamic cycle illustrating the conceptual decomposition of one electron reduction of a quinone Q in water (top reaction) into the desolvation of Q,, its reduction in the gas phase, and the solvation of the resulting Q anion

Instead of calculating the individual solvation/desolvation energies, however, we use MD methods to calculate their difference, $\Delta G^{o}_{hyd}(Q^{-}) - \Delta G^{o}_{hyd}(Q)$.

To find the most accurate method for estimating $\Delta G^{o}_{red}(g)$, our group tested a number of *ab initio* MO, DF, and HF/DF methods(35) and discovered that the "B3LYP" HF/DF method(36) with the 6-311G(3d,p) or 6-311G(d,p) basis sets gives particularly accurate electron affinities for a variety of p-benzoquinones.(18,21-25,33,35,37) Table II compares experimentally measured electron affinities(38-42) with electron affinities calculated for a variety of p-benzoquinones using the B3LYP/6-311G(3d,p) or B3LYP/6-311G(d,p) methods. The average absolute difference between calculated and experimental electron affinities (excluding those with largest experimental error, estimated from UV-Vis spectra, and those calculated within the experimental range of

Table II Calculated(18,21-25,33,35,37) and experimental(38-42) adiabatic electron affinities (EA) for a number of *p*-benzoquinones. Calculations were done using the B3LYP/6-311G(3d,p) or B3LYP/6-311G(d,p) methods. Average absolute error was calculated excluding quinones with an error range or experimental errors of 0.15eV.

Molecule	Calculated EA (eV)	Exptl. EA (eV±0.01)
9,10-anthraquinone	1.56	1.59
2,3-dimethyl-1,4-naphthoquinone	1.63	N/A
p-duroquinone	1.63	1.62
Trimethyl-*p*-benzoquinone	1.66	1.63
2-methyl-1,4-naphthoquinone	1.69	1.67-1.74
2,5-dimethyl-*p*-benzoquinone	1.69	1.76
2,6-dimethyl-*p*-benzoquinone	1.70	1.77
2,3-dimethyl-*p*-benzoquinone	1.74	N/A
1,4-naphthoquinone	1.75	1.73-1.81
Methyl-*p*-benzoquinone	1.77	1.79
p-benzoquinone	1.85	1.91
Chloro-*p*-benzoquinone	2.19	2.05±0.15
2,3-dichloro-*p*-benzoquinone	2.40	2.19±0.15
2,5-dichloro-*p*-benzoquinone	2.48	2.44
2,6-dichloro-*p*-benzoquinone	2.48	2.48
Trichloro-*p*-benzoquinone	2.67	2.56
p-fluoranil	2.62	2.70
p-chloranil	2.83	2.78
2,3-dichloro-5,6-dicyano-*p*-benzoquinone	3.65	N/A
Average absolute error	0.05 eV	
Ubiquinone-1 (model)	1.81	1.86 (UQ-0)
Plastoquinone-1 (model)	1.75	N/A
Menaquinone-1 (model)	1.68	N/A

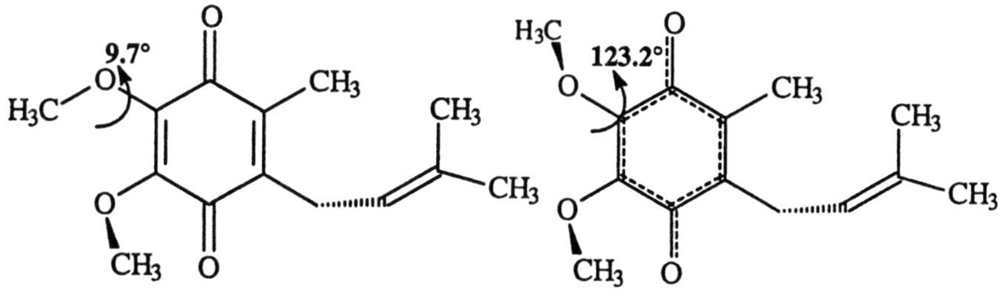

Figure 5. Calculated structures(20) of (a) a model for ubiquinone-1 and (b) its semiquinone anion showing that one methoxy group's conformation changes significantly upon reduction. The adiabatic electron affinity for the model ubiquinone-1 is 1.81 eV(18,20,46) and its electron affinity with the methoxy torsional angle constrained to 9.7° is 1.66 eV.(18)

values) is only 0.05 eV, smaller than experimental error. Except for *p*-fluoranil, calculated electron affinities for all *p*-benzoquinones tested also fall in the correct relative order. Since experimental electron affinities for those marked "N/A" were to our knowledge unavailable at the time they were published, calculated values in the table are predictions. Particularly noteworthy is the good agreement between ubiquinone-1's calculated electron affinity (1.81 eV)(18) and the measured electron affinity for ubiquinone-0 (1.86 ± 0.01 eV), as well as the predicted electron affinities for plastoquinone-1 and menaquinone-1. Also noteworthy is the dependence of ubiquinone's electron affinity on the conformation of its methoxy groups. The effect on thermodynamics has been noted before(43-45) and proposed as a way to modulate the molecule's reduction potential. Figure 5 indicates that the out-of-plane methoxy group is capable of modulating ubiquinone-1's electron affinity over a range of at least 0.15 eV, from 1.66 eV to 1.81 eV (its adiabatic electron affinity).(18) This observation, the thermodynamic cycle of Figure 4 relating electron affinity to reduction free energy, and the Marcus expression relating electron transfer activation energies to free energies ($\Delta G^{\ddagger} = (\Delta G^{\circ} + \lambda_i)/4\lambda_i$, where λ_i is the "reorganization energy"(47)) led Scott Boesch to propose for the first time that the conformation of one ubiquinone methoxy group provides a handle for "torsional tuning" of electron transfer *rates* involving ubiquinone molecules in proteins.(18) In fact, the *Rb. sphaeroides* RC can modulate its electron transfer reactions by controlling methoxy orientations of the primary ubiquinone (which functions as both an electron acceptor and donor) and the secondary ubiquinone acceptor.

The thermodynamic cycle shown in Figure 4 demonstrates that the reduction free energies for *p*-benzoquinones in water are $\Delta G^{\circ}_{red}(aq) =$

$\Delta G^{o}_{red}(g) + (\Delta G^{o}_{hyd}(Q^-) - \Delta G^{o}_{hyd}(Q))$. Since entropic contributions to gas-phase reduction free energies are small, approximately 0.04 eV,(38,40) we use calculated electron affinities displayed in Table II as estimates for the gas-phase free energy difference $\Delta G^{o}_{red}(g)$. The differences in hydration free energies between p-benzosemiquinone anions and the parent p-benzoquinones $(\Delta G^{o}_{hyd}(Q^-) - \Delta G^{o}_{hyd}(Q))$ were estimated by using free energy perturbation./molecular dynamics simulations. To calculate the contribution of hydration free energy differences to the reduction free energy, we used B3LYP-derived force field parameters—structures, vibrational force constants, and atomic charges(16-18,20-25)—in free energy perturbation/MD simulations. The derived force fields were actually used in two separate free-energy perturbation simulations, one in water and one in the gas-phase. These free energies characterize the horizontal arrows in Figure 4 and their difference is equal to the difference in hydration free energies (which characterize the vertical arrows in Figure 4) Results are summarized in Table III and show that the force fields and simulation protocol outlined above give reduction potentials for the series of p-benzoquinones studied with an average absolute error of 0.06 eV. In fact, calculated reduction free energies for p-benzoquinones in water are as close as 10 meV to experiment.(21,22,25,33,37,46)

Table III. Calculated(21,22,25,33,37,46) and experimental(43,48-50) absolute one-electron reduction potentials (eV) for a number of p-benzoquinones in water.

Molecule	Calculated E° (eV)	Exptl. E° (eV±0.06)
p-duroquinone	3.99	4.18
Ubiquinone-1 (model)	4.16	4.20
Trimethyl-p-benzoquinone	4.11	4.28
Plastoquinone-1 (model)	4.22	4.28
p-benzoquinone	4.63	4.54
Chloro-p-benzoquinone	4.62	-----
2,5-dichloro-p-benzoquinone	4.66	4.65
2,3-dichloro-p-benzoquinone	4.70	-----
2,6-dichloro-p-benzoquinone	4.72	4.66
Trichloro-p-benzoquinone	4.80	-----
p-chloranil	4.89	4.78
Average absolute error	0.06 eV	

Ubiquinone Binding in the Photosynthetic Reaction Center of *Rhodobacter sphaeroides*

We recently reported a series of molecular dynamics simulations designed to investigate the preferred binding site(s) of UQ_B in the photosynthetic reaction center for *Rb. sphaeroides*.(51) Simulations of ubiquinone-10 used MD force fields derived from quantum chemical calculations in the same way as those which gave the reduction potentials reported in Table III. Simulations confirmed that UQ_B can bind in the dark adapted site and in the light adapted site occupied by UQ_B^- in the structure of Stowell et al. and illustrated in Figure 6. Simulations also show that UQ_B can migrate from the dark adapted site to the light adapted site, but only under a particularly restrictive set of conditions. First, UQ_B must adopt an orientation with its isoprenyl side chain away from the Phe L216 side chain shown Plate 1 and with its head group oriented to form a hydrogen bond to the His L190 side chain. In addition, the side chain of Glu L212 must be located where it cannot compete with UQ_B for a hydrogen-bond to the His L190 side chain. If any of these conditions are violated, UQ_B remains in the dark adapted site. These simulations imply that Glu L212 acts as a protein conformational gate controlling UQ_B migration to the putative electron transfer active site. Thus, the Glu L212 side chain may also provide a conformational gate for electron transfer from UQ_A^- to UQ_B in the bacterial photosynthetic RCs.

Summary

Work reviewed here shows that hybrid Hartree-Fock/density functional calculations can provide molecular structures of unprecedented accuracy and vibrational frequencies among the most accurate currently available from quantum chemical methods.(18,20-25) Electron affinities calculated by using the B3LYP/6-311G(3d,p) or B3LYP/6-311G(d,p) methods also show unprecedented accuracy,(18,21-25,33,35,37) albeit for a restricted class of molecules. The reported accuracy for the electron affinities of *p*-benzoquinones is not generally expected(52,53) and may result because the added electron is delocalized, so the B3LYP method captures the majority of the electron self-interaction energy for these anions. Electron affinities were predicted for ubiquinone-1, plastoquinone-1, and menaquinone-1 with an accuracy approaching 0.05 eV. In addition, the variation of ubiquinone's calculated electron affinity as the torsional angle of its methoxy groups is varied was proposed to provide a mechanism for the torsional tuning of electron transfer *rates* by proteins. Force fields derived from B3LYP calculations also allow calculation of one-electron reduction potentials for *p*-benzoquinones in water with an average absolute difference of 0.06 eV from experimental

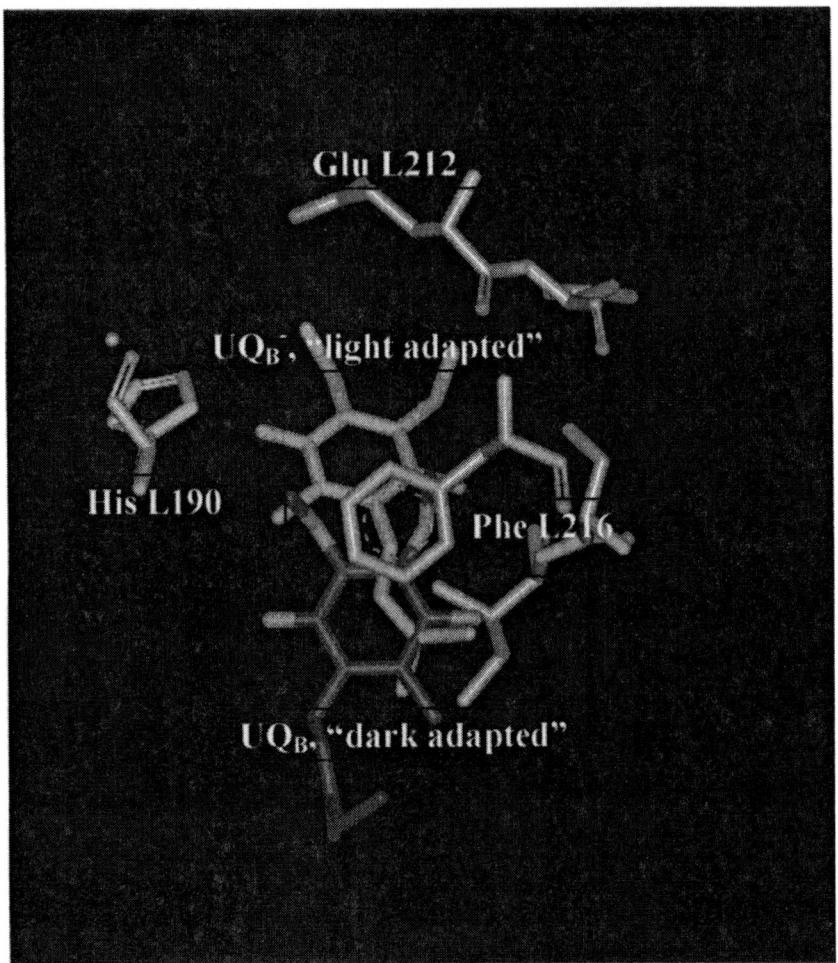

Figure 6. Amino acid side chains near the secondary ubiquinone binding site in the light adapted structure (green) of the Rb. sphaeroides RC. One oxygen of ubiquinone is within hydrogen bonding distance of the His L190 side chain and the Glu L212 side chain is oriented further away. Also shown is the binding site of UQ_B in the dark adapted structure (blue) and the Phe L216 side chain (Created from coordinates reported in reference 14. Isoprenyl chains of the ubiquinones have been truncated for clarity).

(See page 2 in color insert.)

measurements.(21,22,25,33,37,46) Force fields derived in a similar way have been used to model the binding site(s) of the secondary ubiquinone-10 in the photosynthetic reaction center from the bacterium *Rhodobacter sphaeroides*.(51) Not only must ubiquinone-10 adopt a specific orientation to migrate from the electron transfer inactive site to the putative electron transfer active site, but the side chain of Glu L212 must be oriented so it cannot hydrogen bond to the side chain of His L190. Thus, the Glu L212 side chain competes with ubiquinone-10 for hydrogen bonding to the His L190 side chain and Glu L212 is therefore proposed to provide a protein conformational gate controlling ubiquinone-10 migration, and perhaps electron transfer, in the *Rb. sphaeroides* photosynthetic reaction center.

As computer power and computational algorithms advance, work described here will certainly be improved upon. For example, combined quantum mechanical-molecular dynamics studies of molecular solvation and protein-ligand interactions are becoming more common and may soon become routine for thermochemical calculations, as well as binding site studies, of the sort described here.

Acknowledgments

I am especially grateful to the extremely talented co-workers who contributed both directly and indirectly to the work described in this review. Many of them are cited in the references and include Scott E. Boesch, Dr. A. Kurt Grafton, Kevin S. Raymond, Dr. Susan E. Walden, and Dr. Kristopher E. Wise. We are grateful for financial support from the U.S. Department of Energy through grant numbers DE-FG03-97ER14806 and DE-FG03-01ER15164, and from the Oklahoma Center for the Advancement of Science and Technology through OCAST award numbers HN3-011 and H97-091. We also thank the National Science Foundation for generous grants of supercomputer time through award number MCA96-N019.

References

1) Stryer, L. *Biochemistry*; W.H. Freeman: New York, 1995, Chapter 26.
2) Garab, G. *Photosynthesis: Mechanisms and Effects*; Kluwer: Dordrecht, 1998, Vol. I-V; Vol. I-V.
3) Blankenship, R. E.; Madigan, M. T.; Bauer, C. E. *Anoxygenic Photosynthetic Bacteria*; Kluwer: Dordrecht, 1995.

4) Rhee, K.-H. "Photosystem II: the solid structural era,"*Annu. Rev. Biophys. Biomolec. Struc.* **2001**, *30*, 307-328.
5) Okamura, M. Y.; Feher, G. "Proton Transfer in Reaction Centers from Photosynthetic Bacteria,"*Annu. Rev. Biochem.* **1992**, *61*, 861-896.
6) Michel, H.; Deisenhofer, J. "Relevance of the Photosynthetic Reaction Center from Purple Bacteria to the Structure of Photosystem II,"*Biochemistry* **1988**, *27*, 1.
7) Diner, B. A.; Petrouleas, V.; Wendoloski, J. J. "The Iron-Quinone Electron-Acceptor Complex of Photosystem II,"*Physiol. Plant.* **1991**, *81*, 423-436.
8) Feher, G.; Allen, J. P.; Okamura, M. Y.; Rees, D. C. "Structure and Function of Bacterial Photosynthetic Reaction Centres,"*Nature* **1989**, *339*, 111-116.
9) Allen, J. P.; Feher, G.; Yeates, T. O.; Komiya, H.; Rees, D. C. "Structure of the Reaction Center From *Rhodobacter Sphaeroides* R-26: The Cofactors,"*Proc. Natl. Acad. Sci. USA* **1987**, *84*, 5730-5734.
10) Allen, J. P.; Feher, G.; Yeates, T. O.; Komiya, H.; Rees, D. C. "Structure of the Reaction Center from *Rhodobacter Sphaeroides* R-26: Protein-Cofactor (Quinones and Fe^{2+}) Interactions,"*Proc. Natl. Acad. Sci. USA* **1988**, *85*, 8487-8491.
11) Deisenhofer, J.; Michel, H. "The Photosynthetic Reaction Centre From the Purple Bacterium Rhodopseudomonas viridis,"*EMBO J.* **1989**, *8*, 2149-2170.
12) Lancaster, C. R. D.; Michel, H. "Three-dimensional structures of photosynthetic reaction centers,"*Photosynth. Res.* **1996**, *48*, 65-74.
13) Lancaster, C. R. D.; Ermler, U.; Michel, H. "The Structures of Photosynthetic Reaction Centers from Purple Bacteria as Revealed by X-ray Crystallography," in *Anoxygenic Photosynthetic Bacteria*; Blankenship, R. E., Madigan, M. T. and Bauer, C. E., Ed.; Kluwer: Dordrecht, 1995, pp 503-526 and references therein.
14) Stowell, M. H. B.; McPhillips, T. M.; Rees, D. C.; Soltis, S. M.; Abresch, E.; Feher, G. "Light-Induced Structural Changes in Photosynthetic Reaction Center: Implications for Mechanism of Electron-Proton Transfer,"*Science* **1997**, *276*, 812-816.
15) Wheeler, R. A. "Quinones and quinoidal radicals in photosynthesis,"*Theoret. Comput. Chem.* **2001**, *9*, 655-690.
16) Boesch, S. E.; Wheeler, R. A. "π-Donor Substituent Effects on Calculated Structures and Vibrational Frequencies of *p*-Benzoquinone, *p*-Fluoranil, and *p*-Chloranil,"*J. Phys. Chem.* **1995**, *99*, 8125-8134.
17) Boesch, S. E.; Wheeler, R. A. "π-Donor Substituent Effects on Calculated Structures, Spin Properties, and Vibrations of Radical Anions of *p*-Chloranil, *p*-Fluoranil, and *p*-Benzoquinone,"*J. Phys. Chem. A* **1997**, *101*, 8351-8359.
18) Boesch, S. E. *Structures, Bonding, and Properties of Substituted Quinones and Semiquinone Radical Anions*; MS Thesis; University of Oklahoma: Norman, 1996, pp 81.
19) Hagen, K.; Hedberg, K. "Reinvestigation of the molecular structure of gaseous *p*-benzoquinone by electron diffraction,"*J. Chem. Phys.* **1973**, *59*, 158-162.

20) Boesch, S. E.; Wheeler, R. A. "Structures and Properties of Ubiquinone-1 and Its Radical Anion from Hybrid Hartree-Fock/Density Functional Studies,"*J. Phys. Chem. A* **1997**, *101*, 5799-5804.

21) Wise, K. E.; Grafton, A. K.; Wheeler, R. A. "Predicted Structures and Properties of Models for Plastoquinone-1 and its Radical Anion,"*J. Phys. Chem. A* **1997**, *101*, 1160-1165.

22) Wise, K. E. *Computational Studies of Biological Electron Transfer Cofactors and Electron Donor-Acceptor Complexes*; PhD Dissertation; University of Oklahoma: Norman, OK, 1999, pp 203.

23) Grafton, A. K.; Wheeler, R. A. "Structures and Properties of Vitamin K and its Radical Anion Predicted by a Hybrid Hartree-Fock/Density Functional Method,"*J. Mol. Struc. (Theochem)* **1997**, *392*, 1-11.

24) Grafton, A. K.; Wheeler, R. A. "A Comparison of the Properties of Various Fused-Ring Quinones and Their Radical Anions Using Hartree-Fock and Hybrid Hartree-Fock/Density Functional Methods,"*J. Phys. Chem. A* **1997**, *101*, 7154-7166.

25) Grafton, A. K. *A Computational Study of the Structural, Spectroscopic, Thermochemical, and Binding Properties of Quinones involved in Photosynthesis*; PhD Dissertation; University of Oklahoma: Norman, OK, 1998, pp 210.

26) Becker, E. D.; Charney, E.; Anno, T. "Molecular vibrations of quinones. VI. A vibrational assignment for p-benzoquinone and six isotopic derivatives. Thermodynamic functions of p-benzoquinone,"*J. Chem. Phys.* **1965**, *42*, 942-949.

27) Palmo, K.; Pietila, L.-O.; Mannfors, B. "Raman scattering from p-benzoquinone,"*J. Mol. Spectrosc.* **1983**, *100*, 368-376.

28) Trommsdorff, H. P.; Wiersma, D. A.; Zelsmann, H. R. "Vapor-solvent shift of the lowest-frequency vibration of p-benzoquinone and toluquinone and the consequences for the vibrational and electronic spectral assignments,"*J. Chem. Phys.* **1985**, *82*, 48-52.

29) Scott, A. P.; Radom, L. "Harmonic Vibrational Frequencies: An Evaluation of Hartree-Fock, Moeller-Plesset, Quadratic Configuration Interaction, Density Functional Theory, and Semiempirical Scale Factors,"*J. Chem. Phys.* **1996**, *100*, 16502-16513.

30) Rauhut, G.; Pulay, P. "Transferable Scaling Factors for Density Functional Derived Vibrational Force Fields,"*J. Phys. Chem.* **1995**, *99*, 3093-3100.

31) Nonella, M.; Tavan, P. "An unscaled quantum mechanical harmonic force field for p-benzoquinone,"*Chem. Phys.* **1995**, *199*, 19-32.

32) Nonella, M. "A re-investigation of the v_2 and v_3 modes of 1,4-benzoquinone on the basis of a quantum mechanical harmonic force field,"*Chem. Phys. Lett.* **1997**, *280*, 91-94.

33) Wheeler, R. A. "A Method for Computing One-Electron Reduction Potentials and its Application to p-Benzoquinone in Water at 300K,"*J. Am. Chem. Soc.* **1994**, *116*, 11048-11051.

34) Allen, M. P.; Tildesley, D. J. *Computer Simulation of Liquids*; Clarendon Press: Oxford, 1987.
35) Boesch, S. E.; Grafton, A. K.; Wheeler, R. A. "Electron Affinities of Substituted *p*-Benzoquinones from Hybrid Hartree-Fock/Density-Functional Calculations,"*J. Phys. Chem.* **1996**, *100*, 10083-10087.
36) Stephens, P. J.; Devlin, F. J.; Chablowski, C. F.; Frisch, M. J. "Ab Initio Calculation of Vibrational Absorption and Circular Dichroism Spectra Using Density Functional Force Fields,"*J. Phys. Chem.* **1994**, *98*, 11623-11627.
37) Raymond, K. S.; Grafton, A. K.; Wheeler, R. A. "Calculated One-Electron Reduction Potentials and Solvation Structures for Selected *p*-Benzoquinones in Water,"*J. Phys. Chem. B* **1997**, *101*, 623-631.
38) Kebarle, P.; Chowdhury, S. "Electron Affinities and Electron-Transfer Reactions,"*Chem. Rev.* **1987**, *87*, 513-534 and references therein.
39) Heinis, T.; Chowdhury, S.; Scott, S. L.; Kebarle, P. "Electron Affinities of Benzo-, Naphtho-, and Anthraquinones Determined from Gas-Phase Equilibria Measurements,"*J. Am. Chem. Soc.* **1988**, *110*, 400-407.
40) Chowdhury, S.; Grimsrud, E. P.; Kebarle, P. "Entropy Changes and Electron Affinities from Gas-Phase Electron-Transfer Equilibria: $A^- + B = A + B^-$,"*J. Phys. Chem.* **1986**, *90*, 2747-2752.
41) Fukuda, E. K.; McIver, R. T. "Relative electron affinities of substituted benzophenones, nitrobenzenes, and quinones,"*J. Am. Chem. Soc.* **1985**, *107*, 2291-2296.
42) Chen, E. C. M.; Wentworth, W. E. "Comparison of experimental determinations of electron affinities of pi charge transfer complex acceptors,"*J. Chem. Phys.* **1975**, *63*, 3183-3191.
43) Prince, R. C.; Dutton, P. L.; Bruce, J. M. "Menaquinones and Plastoquinones in Aprotic Solvents,"*FEBS Lett.* **1983**, *160*, 273-276.
44) Robinson, H. H.; Kahn, S. D. "Interplay of Substituent Conformation and Electron Affinity in Quinone Models of Quinone Reductases,"*J. Am. Chem. Soc.* **1990**, *112*, 4728-4731.
45) Silverman, J.; Stam-Thole, I.; Stam, C. H. "The Crystal and Molecular Structure of 2-Methyl-4,5-dimethoxy-*p*-quinone (Fumigatin Methyl Ether), $C_9H_{10}O_4$,"*Acta Crystallogr. B* **1971**, *27*, 1846-1851.
46) Grafton, A. K.; Wheeler, R. A. "Amino Acid Protonation States Determine Binding Sites of the Secondary Ubiquinone and its Anion in the *Rhodobacter sphaeroides* Photosynthetic Reaction Center,"*J. Phys. Chem. B* **1999**, *103*, 5380-5387.
47) Marcus, R. A.; Sutin, N. "Electron Transfers in Chemistry and Biology,"*Biochim. Biophys. Acta* **1985**, *811*, 265-322.
48) Rich, P. R.; Bendall, D. S. "The Kinetics and Thermodynamics of the Reduction of Cytochrome c by Substituted *p*-Benzoquinols in Solution,"*Biochim. Biophys. Acta* **1980**, *592*, 506-518.

49) Swallow, A. J. "Physical Chemistry of Semiquinones," in *Function of Quinones in Energy Conserving Systems*; Trumpower, B. L., Ed.; Academic: New York, 1982, pp 59-72.

50) Prince, R. C.; Gunner, M. R.; Dutton, P. L. "Quinones of Value to Electron-Transfer Studies: Oxidation-Reduction Potentials of the First Reduction Step in an Aprotic Solvent," in *Function of Quinones in Energy Conserving Systems*; Trumpower, B. L., Ed.; Academic: New York, 1982, pp 29-33.

51) Walden, S. E.; Wheeler, R. A. "Protein Conformational Gate Controlling Binding Site Preference and Migration of Ubiquinone-B in the Photosynthetic Reaction Center of *Rhodobacter sphaeroides*," *J. Phys. Chem. B* **2002**, *106*, 3001-3006.

52) Raymond, K. S.; Wheeler, R. A. "Compatibility of correlation-consistent basis sets with a hybrid Hartree-Fock/density functional method," *J. Comput. Chem.* **1999**, *20*, 207-216.

53) Rienstra-Kiracofe, J. C.; Tschumper, G. S.; Schaefer, H. F.; Nandi, S.; Ellison, G. B. "Atomic and molecular electron affinities: Photoelectron experiments and theoretical computations," *Chem. Rev.* **2002**, *102*, 231-282.

Chapter 5

Problems Evaluating Energetics of Electron Transfer from Q_A to Q_B: The Light-Exposed and Dark-Adapted Bacterial Reaction Center

Björn Rabenstein and Ernst-Walter Knapp*

Department of Biology, Chemistry and Pharmacy, Institute of Chemistry, Free University of Berlin, Takustrasse 6, D–14195 Berlin, Germany

The energetics of the electron transfer (ET) process from $Q_A^{\bullet -}$ to Q_B in the bacterial photosynthetic reaction center (bRC) is investigated by solving the linear Poisson-Boltzmann equation to calculate the electrostatic energies. The equilibrium distribution of redox states and protonation pattern of all titratable groups in the protein complex was obtained with a Monte Carlo sampling technique. Considering the dark-adapted structure of the bRC from *Rhodobacter sphaeroides*, we found that the ET reaction is uphill in energy by 157 meV, whereas for the light-exposed structure from *Rhodobacter sphaeroides* and for the new structure from *Rhodopseudomonas viridis*, we found that the ET process is downhill in energy by –56 meV and –169 meV, respectively. Implications of a comparison of our results with experiments are discussed. Reasons are given, why previous attempts to compute these energies using a single crystal structure failed.

© 2004 American Chemical Society

Introduction

The bacterial photosynthetic reaction center (bRC) is a transmembrane protein complex of purple bacteria, which converts light energy into electrochemical energy. Since the first crystal structure of the bRC from *Rhodopseudomonas (Rps.) viridis* became available (*1, 2*), the understanding of its function has considerably increased. Nevertheless, many puzzles remain to be assembled. Among them is the question of how the protein environment manages to tune the redox potentials of the cofactors and the pK_a of titratable groups such that the bRC can perform its function.

The central part of the bRC consists of three polypeptide chains (L, M, H), in which ten cofactors are arranged in two branches A and B related by a C_2 symmetry. Electrons are transferred exclusively via the active A-branch, whose cofactors are mainly embedded in the L-chain. The cofactors are one carotenoid (*Rps. viridis* only), one non-heme iron, four bacteriochlorophylls (BC), two of them forming the special pair, two bacteriopheophytins (BP), and two quinones (Q_A, Q_B), which are both ubiquinones (UQ) for the bRC from *Rhodobacter (Rb.) sphaeroides*. In *Rps. viridis* the quinone of the active A branch Q_A is a menaquinone (MQ).

The primary events of photosynthesis in bRCs start in the antenna system, where the light energy is converted to electronic excitation and funneled to the special pair. There a charge separation process occurs and an electron moves in about 2.5 ps to the BP of the A-branch and subsequently in about 200 ps to the quinone Q_A. This electron reduces the quinone Q_B in the millisecond time regime. After this first electron transfer (ET) process from $Q_A^{\bullet-}$ to Q_B, $Q_B^{\bullet-}$ takes up one proton, a second electron is transferred, forming $Q_BH^{\bullet}$, which gets protonated again to form dihydroquinone Q_BH_2. Finally, the dihydroquinone leaves its binding site and is replaced by another oxidized UQ from the quinone pool.

In the first crystal structure of the bRC from *Rps. viridis* (*1, 2*), the Q_B binding site was occupied to 30% only. Now, a new crystal structure with the same resolution is available, which does not have this deficiency (*3, 4*). Also for the bRC from *Rb. sphaeroides* several structures were resolved by X-ray crystallography (*5-10*). While the two crystal structures of the bRC from *Rps. viridis* are very similar and do not differ much even in the neighborhood of the Q_B binding site (*11*), crystals of the bRC from *Rb. sphaeroides* exhibit two structures obtained in the absence (dark-adapted) or presence (light-exposed) of light, which differ significantly at the Q_B binding site (*10*). The Q_B binding site of the dark-adapted crystal structure (PDB code 1aij) from Stowell et al. (*10*) is only similar to the structure obtained by Ermler et al. (*8*). All other X-ray structures of the bRC from *Rb. sphaeroides* resemble more the light-exposed structure. In the light-exposed structure (PDB code 1aig) the Q_B is shifted by 4.5Å towards the cytoplasm side with an accompanying 180° propeller twist about the isoprene tail. For a comparison of the Q_B binding site of the bRC of the dark-adapted and light-exposed crystal structures from *Rb. sphaeroides* (*3, 4*) and *Rps. viridis* (*1, 2*) see Figure 1.

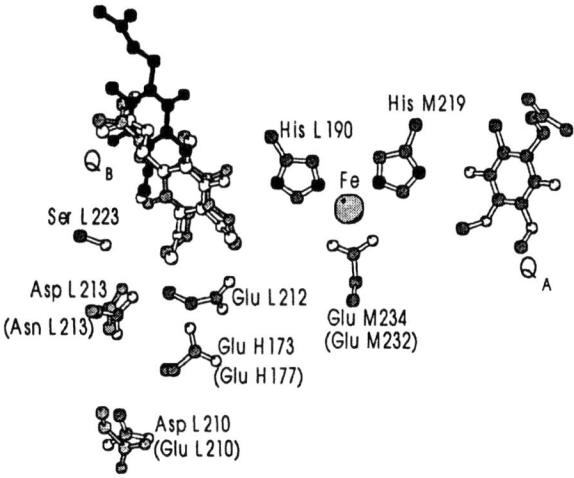

Figure 1. Q_B binding site of different bRC. The structures were superimposed with the Kabsch algorithm (12) considering all displayed molecular components except the two quinones and residue L210. The Q_B of the dark-adapted and light-exposed crystal structures from Rb. sphaeroides (10) are displayed in black and white, respectively. Q_B from Rps. viridis (3, 4) is depicted in gray. Residues common to all three structures, are taken from the light-exposed structure of Rb. sphaeroides. Further residues from Rps. viridis are depicted in gray and denoted in brackets. (Drawn with Molscript (61).)

Previous Computations

In previous studies, we investigated different ET, protein-protein association and cofactor binding reactions of photosynthetic proteins using various theoretical methods (*13-21*). Here, we focus on the ET process from $Q_A^{\bullet}$ to Q_B in the dark-adapted and light-exposed bRC from *Rb. sphaeroides*. It was found that the rate of this ET reaction diminished by several orders of magnitude (*22*) if the bRC from *Rb. sphaeroides* was frozen in the dark-adapted state as compared to the bRC frozen under illumination. This let relatively early to the conclusion that the structures of the light-exposed and dark-adapted bRC should differ and that the bRC has to undergo a conformational transition before the electron can be transferred efficiently from $Q_A^{\bullet-}$ to Q_B (23, 24). In case this conformational change is much slower than the ET reaction, it is gating the transfer process (25). Processes governed by such a conformational gating mechanism were found in reactions of other photosynthetic systems too (26). Since the ET rate from $Q_A^{\bullet-}$ to Q_B did not show a dependence on the free energy difference between the initial ($Q_A^{\bullet-} Q_B$) and final ($Q_A Q_B^{\bullet-}$) state of this reaction, it was concluded that this ET reaction is subject to conformational gating (27, 28).

Recently, Grafton and Wheeler (29) performed molecular dynamics simulations on the Q_B binding site of the dark-adapted and light-exposed structure of the bRC from *Rb. sphaeroides* considering different protonation patterns. They found that the light-exposed structure with $Q_B^{\bullet-}$ requires a protonated Glu L212 and Asp L213, whereas the dark-adapted structure with Q_B^0 is stable only if Glu L212 is protonated and Asp L213 is not.

Based on the crystal structure of the bRC from *Rb. sphaeroides* the energetics of the ET reaction from $Q_A^{\bullet-}$ to Q_B was calculated before, by solving the linear Poisson-Boltzmann (LPB) equation (30). According to these computations, the ET reaction from $Q_A^{\bullet-}$ to Q_B was uphill in energy by 170 meV instead of being downhill by about -52 to -78 meV, as measured experimentally (23, 24, 31, 32). Recently, we performed an equivalent computation on the bRCs from *Rps. viridis* (20, 21), where we found that the ET reaction from $Q_A^{\bullet-}$ to Q_B is downhill in energy and at the same time the other protonation and redox reactions involving the Q_B were mainly in agreement with experiments as far as results were available. Although the considered bRC were from different species, the employed methods were the same as in Ref. (30). One major difference between the two approaches was the charge model used for the polypeptides and for the cofactors, which was considered to be responsible for the discrepancy of the results (20).

Also the protonation pattern of the bRC from *Rps. viridis* was calculated before, at that time, however, without evaluating the energetics of the different redox and protonation states of the two quinones (33). Very recently, Alexov & Gunner (34) repeated the computation of the energetics of the ET reaction between the two quinones, considering now the new structures of the dark-adapted and light-exposed bRC from *Rb. sphaeroides*. Although they failed to obtain the energetics of the ET process from $Q_A^{\bullet-}$ to Q_B appropriately using a single structure, they were successful when using a statistical average over different conformations of the Q_B, the amino acid side chains and water conformers, derived from the different crystal structures of the bRC from *Rb.*

sphaeroides and *Rps. viridis* that are available. Their MCCE method (*34*), which can consider many molecular conformations, closely resembles another method that was recently developed by Beroza and Case (*35*).

Using solely one set of coordinates of the bRC from *Rb. sphaeroides*, Alexov & Gunner (*34*) calculated for the ET process from $Q_A^{\bullet-}$ to Q_B an energy difference of +170 meV. As the energy value obtained before (*30*), this was in disagreement with experimental results. Under the same conditions they repeated this computation also for the bRC from *Rps. viridis*. In contrast to our work (*20*), they could not obtain agreement with experiments (*34*). They were using the more recent crystal structure from Lancaster et al. (*3, 4*), which is better defined in the neighborhood of the Q_B binding site as the crystal structure from Deisenhofer et al. (*1, 2*). We used for our computations (*20*) the structure of Deisenhofer et al. (*1, 2*), since the more suitable crystal structure from Lancaster et al. (*3, 4*) was not publicly available at that time. However, we adjusted the conformation of the Q_B binding site (*20*) to account for the structural differences documented by Lancaster et al. (*11*) (see also Figure 2). Accordingly, we rotated Q_B and the carboxyl group of Glu L212 around the axes displayed in Figure 2.

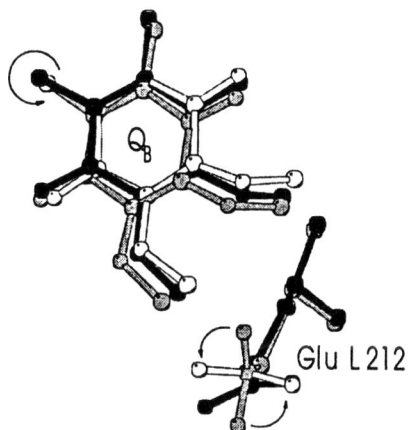

Figure 2. Comparison of bRC structures from Rps. viridis at Q_B binding site. Q_B and Glu L212 of the bRC from Rps. viridis are displayed for the crystal structures from Lancaster at al. (3, 4) (black), from Deisenhofer et al. (1, 2) (gray) and the adjusted structure (white) used previously (20, 21). Arrows indicate the rotation axes used for the structural adjustments. The axes are oriented approximately perpendicular to the drawing plane.

From their results, Alexov & Gunner (34) concluded that the disagreement of the two computations is a result of the differences in the structures that were used. However, besides the differences in the two crystal structures that were used, there were quite a number of other differences between the two computations, as for instance the usage of different charge models for the cofactors and the polypeptide.

Purpose and Methods

In this work, we present calculations on the energetics of the ET process and the corresponding protonation patterns of the bRC from *Rb. sphaeroides* in the dark-adapted and light-exposed state (*10*) and from *Rps. viridis* using the now available crystal structure from Lancaster et al. (*3, 4*). The experimental implications of our results will be discussed. The energies are calculated with the well-established method of continuum electrostatics (*36-42*), which we used successfully before to calculate redox and protonation equilibra in the bRC from *Rps. viridis* (*20, 21*). The dielectric constant within the protein was set to 4 and in the solution to 80. The ionic strength was set to 100 mM. An ion exclusion layer of 2 Å and a solvent probe radius of 1.4 Å were used. We will furthermore investigate reasons for the discrepancies between results from our computations and the ones obtained by Alexov & Gunner (*34*).

Structures

We considered the dark-adapted and light-exposed crystal structures of the bRC from *Rb. sphaeroides*, PDB entry 1aij and 1aig, respectively (*10*). From the two reaction centers in the unit cell, we used the H, L, and M chains with their cofactors. For the computations on the bRC from *Rps. viridis*, we used the new crystal structure from Lancaster et al. (*3, 4*). Here we ignored the cytochrome c subunit, since it is very distant from the quinones. The hydrogen atoms were added with the program CHARMM (*44, 45*) by optimizing the hydrogen positions while fixing all other atoms. All detergent and water molecules were removed. For more explanations see a discussion in Ref. (*20*). In the corresponding cavities, the dielectric constant was set to 80 as outside of the bRC. The influence of a lower dielectric medium at the location of the membrane was ignored, since in the light-exposed structure from *Rb. sphaeroides* the membrane is about 12 Å (10 Å) away from Q_A (Q_B). Except for the cofactors, the two quinones, the BC, and the BP, for which all hydrogen atoms were considered explicitly, we used an extended atom representation, where only polar hydrogen atoms were treated explicitly. The proton of an acidic group -COOH can be bound on either oxygen atom. To avoid this uncertainty, these hydrogen atoms were not explicitly modeled. If an acidic group was protonated, the charge was evenly distributed between the two oxygen atoms. The coordinates of the hydrogen atoms were generated with the program CHARMM (46). The positions of the hydrogen atoms were subsequently optimized by energy minimization, while the positions of the non-hydrogen atoms were fixed.

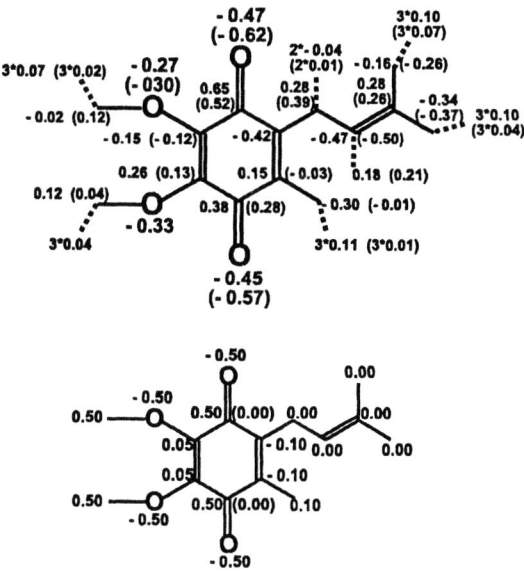

Figure 3. Charges of UQ^0 (UQ^-) used for the bRC. Upper part present work, lower part Alexov & Gunner (34). Circles denote oxygen atoms, dashed lines point to hydrogen atoms. Atoms that are not displayed have charge zero.

Charges

We used the same atomic partial charges as in Ref. (20). The atomic partial charges of the amino acids were taken from the CHARMM parameter set of MSI (46), as far as possible. This set of charges is more complete with respect to amino acids in non-standard protonation states as compared to the CHARMM19 parameter set (44), whose charges are very similar. All other charges were calculated quantum chemically with the program SPARTAN 4.0 (47). The atomic partial charges of the molecular group considered were determined with a CHELPG-like method (48) available in SPARTAN, such that the electrostatic potential is represented properly. The atomic partial charges of the BC and BP were calculated semi-empirically with the PM3 method. The charges of the UQ and MQ in its neutral and negatively charged redox states and of the deprotonated cysteine were calculated ab initio with a 6-31G** basis set. The charges of the phytol chains of the quinones and the BC and BP were set to zero and not considered in the quantum chemical computations. The charges of the quinones are given in Figure 3. The charges of the high-spin non-heme iron and its ligands were calculated by a density functional method (LSDA/VWN) using the DN** basis as implemented in SPARTAN.

Protonation pattern

All computations were performed at a pH value of 7.0. The LPB equation was solved with the program MEAD from Bashford (39) using a grid focusing procedure with a final grid spacing of 0.3 Å. The methodology to generate protonation patterns of proteins by calculating the electrostatic energies from the solution of the LPB equation and by sampling the protonation states with a Monte-Carlo (MC) algorithm was recently reviewed (43). We followed the same procedure as described in Ref. (20). But, instead of using the program MCTI from Beroza et al. (42) for the MC sampling, we used our own development, the program KARLSBERG (49), which possesses some additional features useful for the calculation of the redox potentials of the quinones. Among them are the possibilities to include redox groups, to apply triple MC moves, where the protonation states of three strongly coupled titratable groups are changed simultaneously (21), and to perform biased sampling (30) needed to improve sampling efficiency. The program KARLSBERG is freely available under the GNU general public license from our webserver (http://lie.chemie.fu-berlin/karlsberg/).

Redox Potential of the Quinones

A redox-active group can be treated in the same way as a titratable group. However, the pH dependence of the titratable group is replaced by the external redox potential of the solvent, and the pK_a value is replaced by the redox potential of corresponding redox-active group (43). To avoid the usage of an external redox potential, the quinones are considered as an extended redox-active group, which possesses the redox states (i) $Q_A^- Q_B$ and (f) $Q_A Q_B^-$. For the energy difference between these two states, an external redox potential cancels. A transition between these two states was included in the move set of the MC sampling program KARLSBERG. The free energy difference between the two states i and f is finally calculated from the probabilities $\langle x_i \rangle$ and $\langle x_f \rangle$ to which these two states are occupied according to the relation $\Delta G = -k_B T \ln[\langle x_i \rangle / \langle x_f \rangle]$, where $\langle x_f \rangle = 1 - \langle x_i \rangle$, k_B is the Boltzmann constant, and T the absolute temperature. In case $\langle x_i \rangle$ or $\langle x_f \rangle$ is close to unity, the resulting energy value involves a large statistical error. To avoid this problem, we applied a bias to the bare energies of the two states i and f, such that the values of $\langle x_i \rangle$ and $\langle x_f \rangle$ are close to 0.5 to minimize the statistical error. To calculate the protonation pattern for the pure redox states $Q_A^{\bullet -} Q_B$ or $Q_A Q_B^{\bullet -}$, a large positive or negative bias was applied to force the system to remain completely in one of the two redox states.

Energetics of the Electron Transfer

The value of the free energy of the ET process from Q_A to Q_B, obtained from MC titration of the bRC from *Rb. sphaeroides* at pH 7.0 is –56 meV for the light-exposed and +157 meV for the dark-adapted structure (Table I). Hence, in the dark-adapted bRC the ET process is inhibited in agreement with experiments (*22*). The experimental values of the free energy difference, which are between –78 meV and –52 meV (*23, 24, 31, 32*) at pH values from 6.0 to 8.5, may involve also a contribution from the conformational transition between the dark-adapted and the light-exposed structure. This effect is not contained in our calculated free energy difference of the light-exposed structure.

Table I. Energies and Protonation probabilities for $Q_A^{\bullet -} Q_B$ and $Q_A Q_B^{\bullet -}$

		protonation probability at pH 7.0					
	$Q_A^{\bullet -} Q_B \rightarrow Q_A Q_B^{\bullet -}$	$Q_A^{\bullet -} Q_B$			$Q_A Q_B^{\bullet -}$		
structure	energy	L210	L212	L213	L210	L212	L213
dark, 1aij	+157 meV	0.01	0.27	0.75	0.01	0.60	0.85
light, 1aig	–56 meV	0.02	0.81	0.37	0.00	1.00	0.99
rps. viridis	–169 meV	0.00	0.99	Asn	0.00	1.00	Asn

NOTE: Residue L212 is a glutamate. Residues L210 and L213 are aspartates in bRC from *Rb. sphaeroides*. In bRC from *Rps. viridis* residue L210 is a glutamate and L213 a neutral asparagine (Asn).

Comparing the energy calculated for the light-exposed structure with the corresponding measured values, one may conclude that the contribution from this conforamtional transition should be small. This is corroborated by the following experimental data. (i) At the Q_B binding site the crystal structure of the dark-adapted state exhibits traces, which may come from a small admixture of the light-exposed structure (*10*). (ii) Experiments on the kinetics of the ET process between the two quinones possess two components (*50-52*). The smaller one is an order of magnitude faster than the main component. This can be understood, if the main component of the ET rate is determined by the conformational transition between the dark-adapted and light-exposed states. The experimental value of the energy for the bRC from *Rps. viridis* is −150 meV at pH 7.5 (*51*), which is about 100 meV more negative than the corresponding value of the bRC from *Rb. sphaeroides*. This energy essentially corresponds to the difference in the redox potentials of the quinones MQ and UQ in aqueous solution, which is −699 mV for $MQ/MQ^{\bullet-}$ and −592 mV for $UQ/UQ^{\bullet-}$ compared to standard calomel reference electrode (SCE) (*20*). Hence, though the neighborhood of the quinones in the two reaction centers is not identical, the relative shift in the redox potentials invoked by the protein matrix is the same. To calculate the energy difference of the bRC from *Rps. viridis*, the new structure from Lancaster et al. was used (*3, 4*). The obtained value of 169 meV compares well to our former value of −160 meV was obtained by using the structure from Deisenhofer et al. (*1, 2*), which was adjusted to account for the structural differences at the Q_B binding site (*20*) (see Figure 2). This does not support the assumption from Alexov & Gunner (*34*), that the discrepancies in the computed energies result from the differences in the used structures.

Protonation of Residues near Q_B

According to our computations, L210 carries nearly a unit negative charge for all structures and redox states of the bRC. This is in contrast with a recent computation on the bRC from *Rb. sphaeroides*, where at pH=7 Asp L210 is unprotonated in the redox states $Q_A Q_B$ and $Q_A^{\bullet-} Q_B$, but protonated to 65% in the redox state $Q_A Q_B^{\bullet-}$ (*34*). On the other hand, FTIR measurements on the bRC from *Rb. sphaeroides* exhibited changes of protonation going along with the ET reaction from $Q_A^{\bullet-} Q_B$ to $Q_A Q_B^{\bullet-}$ only at the carboxylic group of Glu L212 (*50, 54*) and it was concluded that Asp L210 is always unprotonated (*54*). In our computations, the protonation probability of Glu L212 in the light-exposed structure of *Rb. sphaeroides* increases upon ET between $Q_A^{\bullet-} Q_B$ and $Q_A Q_B^{\bullet-}$ from 0.83 to 0.99. At the same time, the protonation probability of Asp L213 increases by 0.53, such that the total increase in charge is 0.69. In contrast, FTIR measurements exhibited no change at Asp L213, whereas the protonation at Glu L212 was measured in different labs to increase by 0.3 − 0.6

(*50*) or 0.3 – 0.4 (*54*). However, the individual contributions of the two residues to the total calculated charge increase are relatively uncertain, since these residues are strongly coupled.

Probing the stability of the light-exposed and dark-adapted structures of the bRC by molecular dynamics simulation, Grafton and Wheeler found for the light-exposed structure that both residues L212 and L213 should be protonated, whereas for the dark-adapted structure only Glu L212 should be protonated (*29*). The latter does not fully agree with our computations of the electrostatic energies of protonation pattern, where in the dark-adapted structure Glu L212 is protonated with probability 0.3 only. However, the sum of protons at residues L212 and L213 is about 1.0 for the dark-adapted structure. Since these two residues are strongly coupled, the distribution of the proton over the two residues can be shifted with a small amount of energy. On the other hand, it may also be possible that similar results can be obtained from molecular dynamics simulation using a different distribution of the proton for the dark-adapted structure.

In the bRC from *Rps. viridis*, residue L213 is an asparagine, which is always uncharged, corresponding to a protonated aspartate L213 in the bRC from *Rb. sphaeroides*. Since in FTIR measurement no change in the protonation of carboxylate groups was observed, while the electron was transferred from $Q_A^{\bullet-} Q_B$ to $Q_A Q_B^{\bullet-}$ (*55, 56*) and Glu L212 should at least be protonated in the final redox state $Q_A Q_B^{\bullet-}$, Glu L212 is supposed to be protonated in both redox states $Q_A^{\bullet-} Q_B$ and $Q_A Q_B^{\bullet-}$.

Discrepancies in Electrostatic Energies

There are a number of reasons that can lead to discrepancies of computed electrostatic energies, if different labs are considering the same molecular system. These are:

- Different techniques or degrees of accuracy can be employed to solve the LPB equation.
- Different coordinates can be used for the same molecular system or for a large molecular system different fractions of the system can be used. Water molecules in a protein can be considered explicitly or the dielectric constant in the resulting cavities is set to 80.
- Different charge models (atomic charges, dielectric constant, molecular surface) can be used.

Differences in Solving the LPB Equation

First, we like to discuss influences on the computed electrostatic energies, which are due to the technical procedure by which the LPB equation is solved. A standard procedure to solve the LPB equation, which we also use here, is to place the molecular system in the center of a finite lattice, where electrostatic potential, charges, dielectric constant and ionic strength are discretized. The size of the initially used lattice should be large enough that the electrostatic potential vanishes nearly at the boundary.

In our approach, the initial lattice size was 250 Å as compared to the largest distances between atoms in x, y and z directions, which were 77 Å, 71 Å, and 76 Å for the bRC from *Rb. sphaeroides*, respectively. For such a large lattice the lattice constant was set to 2.5 Å, to avoid exceedingly large memory and CPU requirements. This lattice constant is too large to obtain a faithful representation of the various electrostatic quantities inside of the molecular system, but is sufficient to get reliable values of the electrostatic potential at the boundaries. In a focussing procedure (*57*), these values were used for the boundaries of a second, smaller lattice of 100 Å size and a lattice constant of 1.0 Å. In a third step, an even smaller lattice of 45 Å size and a lattice constant of 0.3 Å, centered at the molecular group considered, was used, where again the electrostatic potential at the boundaries was taken from the preceding computation. This lattice was fine enough to represent the electrostatic potential within the protein faithfully.

In a recent computation of electrostatic energies by Alexov & Gunner (*34*) the lattice constant for the last focusing step was set to the relatively large value of 0.83 Å. This may have been necessary to keep the demand of CPU-time in a reasonable range, since several thousand different protein conformations were considered, which required to solve the LPB equation correspondingly often.

We investigated the influence of the lattice constant used for the final focussing step on the energies by evaluating all necessary electrostatic potentials to calculate the energy of the ET process in the light-exposed structure of the bRC from *Rb. sphaeroides* also for the larger lattice constant of 0.83 Å. All other parameters of our computation in particular also the atomic partial charges were kept at the same values as before. Calculating the energy where the origin of the low resolution lattice was shifted by half of a lattice constant a = 0.415 Å in each Cartesian direction, (a, 0, 0), (0, a, 0), (0, 0, a), and (a, a, 0), (a, 0, a), (0, a, a), and (a, a, a), we obtained energy values, which varied by at most 26 meV from the value obtained by using the 0.3 Å lattice. The average value of the energy obtained with the low resolution lattice was about 0.20 meV lower than the one calculated with the high resolution lattice.

Different Coordinates and Molecular Composition

Since our method to compute electrostatic energies of molecular systems is based on a single crystal structure, the usage of explicit water molecules would introduce arbitrariness due to the unknown coordinates of the hydrogen atoms (*17*). Hence, this possibility is not seriously considered here. The situation changes completely, if many conformations are employed as it is done by Alexov & Gunner and Berosa & Case, Ref. (*34, 35*), such that one can account for different hydrogen bonding schemes involving also water molecules.

An obvious reason to obtain different results is certainly the use of different sets of coordinates. Simply by adding hydrogen atoms to the crystal structure of a protein, different research groups may obtain different coordinates for these atoms. These differences should be small for hydrogen atoms, where the bonding geometry prescribes a unique position, as for instance for the polar hydrogen of the protein backbone. More complex cases are the histidine and the acidic groups of amino acid side chains, which we handle the same way as Bashford et al. (*40*) and Beroza et al. (*30*). The polar hydrogen atom at the imidazole ring of a deprotonated (neutral) histidine can be attached at either of the two imidazole nitrogen atoms. In Monte Carlo sampling of the protonation states both possibilities are considered. The polar hydrogen atom of a protonated acidic group is not represented explicitly. Instead of that its unit positive charge is symmetrically distributed among the two carbonyl oxygen atoms. In computations by Alexov & Gunner (*34*) these protons were modeled explicitly and placed at all possible positions during the Monte Carlo sampling. However, for computations using a single protein conformation Alexov & Gunner (*34*) used the conformation, which according to the Monte Carlo sampling was the most probable one with respect to the ground state, i.e. at pH7.0 and both quinones uncharged ($Q_A^0 Q_B^0$), instead of using the atomic coordinates from a single crystal structure. As a consequence, the decision where to place the proton of an neutral acidic group in a charged redox state as for instance $Q_A^{\bullet-} Q_B$ or $Q_A Q_B^{\bullet-}$ is done for the non-related uncharged state $Q_A^0 Q_B^0$, where the considered acidic group may be most of the time unprotonated such that the next probable protonated state can be quite arbitrary. A symmetric redistribution of the unit charge of the proton over the acidic oxygens can avoid such arbitrariness.

An example for the usage of different sets of coordinates that we are presenting here is the use of two different structures of the bRC from *Rps. viridis*, namely the new structure from Lancaster et al. (3, 4), PDB code 2prc and the older structure from Deisenhofer et al. (1, 2), PDB code 1prc, which we adjusted at the Q_B binding site to match the new structure (20). The resulting energy differences of the ET reaction from $Q_A^{\bullet-} Q_B$ to $Q_A Q_B^{\bullet-}$ at pH 7.5 were –

160 meV using the crystal structure 1prc (20) and −152 meV using structure 2prc, as compared to the experimental value of −150 meV (53). These deviations are probably close to the uncertainty limits of the computational and experimental methods and therefore not significant. This is in contrast to the arguments from Alexov & Gunner (34), which attributed the energy deviations to the differences of the structures used for these computations. However, the variation in the energy value using the two different structures is even smaller than the uncertainty of the reaction energy calculated with the low resolution lattice where the lattice constant was 0.83 Å.

Dielectric Boundary, Dielectric Constant and Atomic Charges

To evaluate electrostatic energies of molecular systems faithfully, the values employed for atomic charges, dielectric constant of the solute and van der Waals radii of the atoms used for the definition of the dielectric boundaries should be determined consistently.

The dielectric boundary between protein and solvent is generated by rolling the center of a sphere with radius 1.4 Å on the van der Waals surface of the protein. The van der Waals surface is defined by the boundary surface of the volume of the protein, which is obtained by merging the volumes of all atoms contained in the protein. The volumes of individual atoms are determined by their van der Waals radii. Locating small and isolated cavities in large solute molecules as for instance in proteins may cause problems. Deciding whether a small cavity in a protein is large enough to host a solvent molecule represented by the sphere of 1.4 Å radius may depend critically on the values of van der Waals radii used. Thus, with a small decrease of appropriate van der Waals radii, an initially empty cavity, where the dielectric constant adopts the same value as in the solute, will be considered to be filled with solvent and adopt the correspondingly larger dielectric constant of the solvent. If such cavities are close to titratable and redox-active groups, the exact treatment of these cavities can have a large influence on the computed electrostatic energies of those groups.

In the work of Alexov & Gunner (34) PARSE charges and van der Waals radii were used (58). The PARSE charges were adjusted to reproduce measured solvation energies of many different small molecular compounds by considering electrostatic energy and a surface energy term of a single solute conformation. For the adjustment the dielectric constant was set to 80 in the solvent and to account for electronic polarizability it was set to 2 within the solute. Since the molecular compounds considered were small, nuclear polarization effects could safely be neglected. Alexov & Gunner (34) used a dielectric constant of 4 for a single protein conformation as well as for averages over many protein conformations.

In our work atomic charges from the CHARMM version of MSI (46) were used. The CHARMM charges were adjusted to represent the interactions within a protein for dynamics simulation using a dielectric constant of unity everywhere. The value of unity for the dielectric constant implies that nuclear and electronic polarization effects are considered explicitly. The later is done by increasing nuclear polarization effects with enlarged values of the atomic charges. However, this effect is not operative, if only a single conformation of the protein is considered. Hence, the dielectric constant within the protein was set to the value 4, if only a single conformation was used (20). The dielectric constant was set to 2, if the protein conformation was energy minimized, while the protonation pattern was determined self-consistently (21), since with this procedure one accounts at least in part for conformational flexibility.

The MSI CHARMM (46) version we used, refers to an extended atom model, where non-polar hydrogen atoms are not treated explicitly. The same is practically the case for the PARSE (58) charge model, where non-polar hydrogen atoms have vanishing charges and van der Waals radii, such that the hydrogen atoms are screened by the atoms where they are attached to. With the exception of the polar hydrogen atom of the protein backbone, whose atomic partial charge is +0.25 and +0.40 in the MSI CHARMM and PARSE parameter set, respectively, the charges of the protein backbone and the non-polar amino acid side chain atoms exhibit only minor differences. However, some of the charges used for the side chains of polar and charged residues and for cofactors exhibit major differences. A general trend is, that PARSE charges are more localized and have larger absolute values than MSI CHARMM charges.

A comparison of the atomic partial charges used for UQ is shown in Figure 3. Also the charges for the non-heme iron complex used here (20, 21) differ considerably from the charges used by Alexov & Gunner (34), where the iron possesses the bare charge of +2 and the ligands, which are two histidines (L190, M219) and an unprotonated glutamate (M234), have the same charges as the corresponding unligated residues.

For the computation of electrostatic energies based on a single conformation, both research labs used the same dielectric constants, i.e. 4 within the protein and 80 in solution and in protein cavities. But it is astonishing to notice that Alexov & Gunner used inside the protein a dielectric constant of 4 also while sampling over many conformations. It is generally accepted that the dielectric constant accounts for dielectric screening due to electronic and nuclear polarization effects, which are not considered explicitly. However, using the a flexible molecular model, nuclear polarization is considered explicitly such that the dielectric constant should be smaller than 4 to account only for the remaining electronic polarization. This is corroborated by a recent application of the LPB equation, where we tried to consider nuclear polarization effects by applying energy minimization. To obtain results, which were equivalent to the treatment without energy minimization, we had to reduce the dielectric constant inside the protein from the value of 4 to 2 (*21*).

Table II. Calculated Energies of the ET $Q_A^{\bullet -} Q_B \rightarrow Q_A Q_B^{\bullet -}$ in *Rps. viridis* Considering Different Atomic Charges

Charges		MQ ≠ UQ		MQ = UQ	
$Q_A Q_B$	Fe	pH 7.0	pH 7.5	pH 7.0	pH 7.5
Rabenstein	Rabenstein	-169 meV	-152 meV	-62 meV	-45 meV
Alexov	Alexov	-116 meV	-104 meV	-9 meV	+3 meV
Rabenstein	Alexov	-204 meV	-186 meV	-97 meV	-79 meV
Alexov	Rabenstein	-86 meV	-70 meV	+21 meV	+37 meV

NOTE: The keywords *Rabenstein* and *Alexov* indicate that charges were taken from Refs. (*20, 21*) and (*34*), respectively.

To investigate the influence of the different atomic partial charges at the quinones and the non-heme iron on the energetics of the ET process, we repeated our calculation of the free energy difference between the redox states $Q_A^{\bullet -} Q_B$ and $Q_A Q_B^{\bullet -}$ in the bRC from *Rps. viridis*, where now the quinone and non-heme iron charges were taken from Alexov & Gunner (*34*) (see Table II). The energy value obtained by Alexov & Gunner (*34*) was positive, i.e. the ET reaction would be uphill in energy, but the exact value was not given. The energy values given in columns 3 and 4 of Table II account properly for the difference of the redox potentials of UQ and MQ in aqueous solution, which is ΔG_{UQ-MQ} = 107 mV (*20*). The values given in column 5 and 6 ignore this difference, as it would be appropriate for the bRC from *Rb. sphaeroides*. They differ exactly by this amount from the corresponding energy values given in column 3 and 4. The calculated energy values with our charge model (*Rabenstein, Rabenstein*) are close to the experimental values, which are –175 meV and –150 meV at pH 6.0 and 7.5 respectively (*53*). Also the trend in the pH-dependence of the energies is reproduced by our computations. As one can see by comparing the results for different charge models, varying the charges at the non-heme iron complex alone has a relatively small effect, though the energy becomes even smaller using the crude charge model from Alexov & Gunner (*34*). The influence of the charges at the quinones is more significant and goes in the right direction, but cannot explain a change in sign of the energy as it was obtained by Alexov & Gunner (*34*). However, a change in the sign of the energy value would be possible, if one ignores the difference in the redox potentials between UQ and MQ as is demonstrated with the energy values given in column 5 and 6 of Table II.

Finally we would like to comment the result, that Beroza et al. (*30*) obtained for the energy of the ET process from $Q_A^{\bullet -} Q_B$ to $Q_A Q_B^{\bullet -}$ in the bRC from *Rb. sphaeroides*, which was also up-hill in energy by +170 meV instead of being down-hill by −52 meV to −78 meV as found in experiments (*23, 24, 31, 32*). Although they used a single protein conformation, they took a different structure of the bRC (PDB code 4rcr) (*6*). The same values of the dielectric constant were used, but they were using a different charge model. The charges of the protein were taken from DISCOVER (*59*), the van der Waals radii were taken from a precursor of the PARSE parameter set (*60*). The charge parameters of DISCOVER correspond to a true all atom model, where also non-polar hydrogen atoms carry +0.1 elementary charge, similar as in the more current CHARMM22 force field (*45*). Consequently, the van der Waals radii of polar and non-polar hydrogen atoms adopt the non-vanishing value of 1.0 Å. The charges from DISCOVER differ considerably from the PARSE and MSI CHARMM charges. The charge model used for the cofactors was however crude. All atoms of the cofactors were considered to carry no charge with the following exceptions. (1) The oxygen and carbon atoms of carbonyl groups and in particular of the quinones in the neutral charge state carry an elementary charge of −0.38 and +0.38, respectively. (2) The negative charge on the quinones was distributed with 1/3 on each of the carbonyl oxygen and 1/3 evenly distributed over the six carbon five ring atoms. (3) From the +2 elementary charge of the non-heme iron +0.7 remained at the iron and +1.3 elementary charge was delocalized evenly over the ring atoms of the four histidines ligating the iron. Hence, the charge model used for the non-heme iron is more detailed and the charge pattern of the reduced quinones is somewhat more realistic than those used in the work of Alexov & Gunner (*34*). Nevertheless, it is not obvious to us, which influences were responsible for the deviations between the reaction energy value for the ET process from $Q_A^{\bullet -} Q_B$ to $Q_A Q_B^{\bullet -}$ calculated by Beroza et al. (*30*) and the experimental value.

Summary of Results on the Electron Transfer

We have investigated the energetics of the ET process from $Q_A^{\bullet -} Q_B$ to $Q_A Q_B^{\bullet -}$ for the light-exposed and dark-adapted structure of the bRC from *Rb. sphaeroides* (*10*) and for the new structure from *Rps. viridis* (*3, 4*). In contrast to recent work (*34*), where many protein conformations were considered in a

Monte Carlo method, we used a single conformation given by the corresponding crystal structure.

We found that for the light-exposed structure the energy difference between the redox states $Q_A^{\bullet -} Q_B$ and $Q_A Q_B^{\bullet -}$ is −56 meV, i.e. the ET process is downhill in energy, whereas for the dark-adapted structure the energy value is +157 meV, i.e. the ET reaction is practically blocked. Both results are in agreement with experiments. The energy value calculated for the light-adapted structure of the bRC from *Rb. sphaeroides* differs from the corresponding energy value of the structure of the bRC from *Rps. viridis* (3, 4) essentially by the difference in the redox potentials of the quinones in the Q_A binding site. In *Rps. viridis* there is a menaquinone (MQ) and in Rb. sphaeroides there is a ubiquinone (UQ), whose difference of the redox potential in aqueous solution was calculated to be ΔG_{UQ-MQ} = 107 meV (20). Thus, although the protein environment at the Q_A and Q_B binding site differ between the bRC from *Rb. sphaeroides* and *Rps. viridis*, the total shifts in the redox potentials of the two quinones are the same.

From our computations on the light-exposed structure of the bRC from *Rb. sphaeroides* it can be concluded that the sum of protons residing on the acidic residues L212 and L213 is about unity in the initial redox state $Q_A^{\bullet -} Q_B$ and about two in the final redox state $Q_A Q_B^{\bullet -}$. In the bRC from *Rps. viridis* Asp L213 is replaced by an electro-neutral asparagine and Glu L212 is fully protonated in both redox states. The later agrees with FTIR measurements, which cannot find a protonation change at a carboxylic group, while the electron is transferred from $Q_A^{\bullet -}$ to Q_B (55, 56). In FTIR experiments on the bRC from *Rb. sphaeroides*, it was found that the protonation probability of Glu L212 increases by about 0.3 – 0.4 (54), if the electron is transferred from $Q_A^{\bullet -}$ to Q_B and that Asp L210 remains unprotonated. That agrees well with our computations, where however part of the increase in protonation is shifted to Asp L213, which is strongly coupled with Glu L212.

Summary of Sources for Discrepancies

We have review extensively different sources, which may result in discrepancies of electrostatic energies in proteins computed by solving the Poisson-Boltzmann equation. They can be belong to three different groups: (i) differences in the technique to solve the LPB equation, (ii) differences in coordinates, (iii) differences in the charge model. The first point is not critical unless serious mistakes are made.

Differences in coordinates may well give rise to significant discrepancies in calculated electrostatic energies. In particular the presence of explicit water molecules or their absence can have an influence on the results that are obtained with a single conformation only (*17*). However, we have demonstrated that the differences between the new crystal structure of the bRC from *Rps. viridis* (*3, 4*) used by Alexov & Gunner (*34*) and the one we used do not change results on the calculated energy of the ET reaction from $Q_A^{\bullet-}Q_B$ to $Q_AQ_B^{\bullet-}$ (*20*). A serious problem is certainly that Alexov & Gunner (*34*) did not use the appropriate crystal structure, when they calculated the energy difference between the redox states $Q_A^{\bullet-}Q_B$ to $Q_AQ_B^{\bullet-}$ with a single conformation only. The conformation they used was rather a complex combination of conformers of the individual residues taken from crystal structures of different bRC, which they obtained as most probable conformation from their Monte Carlo sampling procedure performed in the redox ground state $Q_A^0Q_B^0$ (*34*). Hence, there is a considerable arbitrariness if a single conformation is used. This problem is absent if many conformations are used, which would explain the agreement with experimental data in latter case.

There are a number of problems relating to dielectric boundary and atomic partial charges, which can also have a large influence on the results of calculated electrostatic energies. Different charge models were used for the components of the protein, whose influence cannot easily be estimated. The charge models used for the cofactors were also different. We could show that the simplified charge model used for the quinones can be made responsible for roughly one quarter of the total deviation between our calculated energy value and the one from Alexov & Gunner (*34*). Finally there is the problem with small cavities in proteins, which, depending on the exact values of the van der Waals radii used, may or may not possess the large dielectric constant of the solvent. When such a cavity is close to a redox-active group, this uncertainty can have a strong effect on the value of the redox potential of that group. However, there is no hint that such a cavitiy is close to one of the two quinones. It will require further careful investigations to understand and control the various parameters influencing the results from computations of electrostatic energies of complex molecular systems like proteins.

Acknowledgements

We thank Dr. Donald Bashford and Dr. Martin Karplus for providing the programs MEAD and CHARMM, respectively. We are grateful to Dr. Matthias Ullmann for useful discussions. This work was supported by the Deutsche Forschungsgemeinschaft Sfb 498, project A5.

References

1. Deisenhofer, J.; Epp, O.; Miki, K.; Huber, R.; Michel, H. *Nature* **1985**, *318*, 618-624.
2. Deisenhofer, J.; Epp, O.; Sinning, I.; Michel, H. *J. Mol. Biol.* **1995**, *246*, 429-457.
3. Lancaster, C.R.D.; Michel, H. *Structure* **1997**, *5*, 1339-1359.
4. Lancaster, C.R.D.; Michel, H. *J. Mol. Biol.* **1999**, *5*, 883-898.
5. Allen, J.P.; Feher, G.; Yeates, T.O.; Komiya, H.; Rees, D.C. *Proc. Natl. Acad. Sci. U.S.A.* **1987**, *84*, 6162-6166.
6. Yeates, T.O.; Komiya, H.; Chirino, A.; Rees, D.C.; Allen, J.P.; Feher, G. *Proc. Natl. Acad. Sci. U.S.A.* **1988**, *85*, 7993-7997.
7. Chang, C.H.; El-Kabbani, O.; Riede, D.; Norris, J.; Schiffer, M. *Biochemistry* **1991**, *30*, 5353-5360.
8. Ermler, U.; Fritzsch, G.; Buchanan, S.K.; Michel, H. *Structure* **1994**, *2*, 925-936.
9. Arnouox, B.; Gaucher, J.-F.; Ducruix, A.; Reiss-Husson, F. *Acta. Cryst. D* **1995**, *51*, 368-379.
10. Stowell, M.H.B.; McPhillips, T.M.; Rees, D.C.; Soltis, S.M.; Abresch, E.; Feher, G. *Science* **1997**, *276*, 812-816.
11. Lancaster, C.R.D.; Michel, H. In *The Reaction Center of Photosynthetic Bacteria – Structure and Dynamics*; Michel-Beyerle, M.E., Ed.; Springer: Berlin, 1996; pp23-35.
12. Kabsch, W. *Acta Cryst.* **1976**, *A32*, 922-923.
13. Muegge, I.; Ermler, U.; Fritzsch, G.; Knapp, E.W. *J. Phys. Chem.* **1995**, *99*, 17917-17925.
14. Ullmann, G.M.; Kostic, N.M. *J. Am. Chem. Soc.* **1995**, *117*, 4766-4774.
15. Muegge, I.; Apostolakis, J.; Ermler, U.; Fritzsch, G.; Lubitz, W.: Knapp, E.W. *Biochemistry* **1996**, *35*, 8359-8370.
16. Apostolakis, J.; Muegge, I.; Ermler, U.; Fritzsch, G.; Knapp, E.W. *J. Am. Chem. Soc.* **1996**, *118*, 3743-3752.
17. Ullmann, G.M.; Muegge I.; Knapp E.W. In *The Reaction Center of Photosynthetic Bacteria – Structure and Dynamics*; Michel-Beyerle, M.E., Ed., Springer: Berlin, 1996; pp143-155.
18. Ullmann, G.M.; Knapp, E.W.; Kostic, N.M. *J. Am. Chem. Soc.* **1997**, *119*, 42-52.
19. Ullmann, G.M.; Hauswald, M.; Jensen, A.; Kostic, N.M.; Knapp, E.W. *Biochemistry* **1997**, *36*, 16187-16196.
20. Rabenstein, B.; Ullmann, G.M.; Knapp, E.W. *Biochemistry* **1998**, *37*, 2488-2495.

21. Rabenstein, B.; Ullmann, G.M.; Knapp, E.W. *Eur. Biophys. J.* **1998**, *27*, 626-637.
22. Kleinfeld, D.; Okamura, M.Y.; Feher, G. *Biochemistry* **1984**, *23*, 5780-5786.
23. Kleinfeld, D.; Okamura, M.Y.; Feher, G. *Biochim. Biophys. Acta* **1984**, *766*, 126-140.
24. Mancino, L.J.; Dean, D.P.; Blankenship R.E. *Biochim. Biophys. Acta* **1984**, *764*, 46-54.
25. Davidson, V.C. *Biochemistry* **1996**, *35*, 14035-14039.
26. Zhou, J.S.; Kostic, N.M. *J. Am. Chem. Soc.* **1993**, *115*, 10796-10804.
27. Graige, M.S.; Feher, G.; Okamura, M.Y.. *Biophys. J.* **1996**, *70*, Abstr. A10.
28. Graige, M.S.; Feher, G.; Okamura, M.Y. *Proc. Natl. Acad. Sci. U.S.A.* **1998**, *95*, 11679-11684.
29. Grafton, A.K.; Wheeler, R.A. *J. Chem. Phys.* **1999**, *103*, 5380-5387.
30. Beroza, P.; Fredkin, D.R.; Okamura, M.Y.; Feher, G. *Biophys. J.*, **1995**, *68*, 2233-2250.
31. Arata, H.; Parson, W.W. *Biochim. Biophys. Acta* **1981**, *253*, 187-202.
32. Tandori, J.; Sebban, P.; Michel, H.; Baciou, L. *Biochemistry* **1999**, *38*, 13179-13187.
33. Lancaster, C.R.D.; Michel, H.; Honig, B.H.; Gunner, M.R. *Biophys. J.*, **1996**, *70*, 2469-2492.
34. Alexov, E.G.; Gunner, M.R. *Biochemistry* **1999**, *38*, 8253-8270.
35. Beroza, P.; Case, D.A. *J. Phys. Chem.* **1996**, *100*, 20156-20163.
36. Gilson, M.K.; Honig, B. *Biopolymers* **1986**, *25*, 2097-2119.
37. Nicholls, A.; Honig. B. *J. Comput. Chem.* **1991**, *12*, 435-445.
38. Davis, M.E.; McCammon, J.A. *J. Comput. Chem.* **1991**, *12*, 909-912.
39. Bashford, D.; Karplus, M. *J. Phys. Chem.* **1991**, *95*, 9557-9561.
40. Bashford, D.; Case, D.A.; Dalvit, C.; Tennant, L.; Wright, P.E. *Biochemistry* **1993**, *32*, 8045-8056.
41. Honig, B.; Nicholls, A. *Science* **1995**, *268*, 1144-1149.
42. Beroza, P.; Fredkin, D.R.; Okamura, M.Y.; Feher, G. *Proc. Natl. Acad. Sci. U.S.A.* **1991**, *88*, 5804-5808.
43. Ullmann, G.M.; Knapp, E.W. *Eur. Biophys. J.* **1999**, *28*, 533-551.
44. Brooks, B.R.; Bruccoleri R.E.; Olafson, B.D.; States, D.J.; Swaminathan, S.; Karplus, M. *J. Comp. Chem.* **1983**, *4*, 187-217.
45. MacKerell, Jr., A.D.; Bashford, D.; Bellot, M.; Dunbrack, Jr., R.L.; Evanseck, J.D.; Field, M.J.; Fischer, S.; Gao, J.; Guo, H.; Ha, S.; Joseph-McCarthy, D.; Kuchnir, L.; Kuczera, K.; Lau, F.T.K.; Mattos, C.; Michnick, S.; Ngo, T.; Nguyen, D.T.; Prodholm, B.; Reiher, III, W.E.; Roux, B.; Schlenkrich, M.; Smith, J.C.; Stote, R.; Straub, J.; Watanabe,

M.; Wiórkiewicz-Kuczera, J.; Yin, D.; and Karplus, M. *J. Phys. Chem.* **1998**, *102*, 3586-3616.
46. CHARMm parameters, MSI, Inc., San Diego, CA, **1993**.
47. SPARTAN version 4.0, Wavefunction, Inc., Irvine, CA, **1995**.
48. Breneman, C.N.; Wiberg, K.B. *J. Comput. Chem.* **1990**, *11*, 361-373.
49. Rabenstein, B. *Karlsberg online manual* **1999**, http://lie.chemie.fu-berlin/karlsberg/.
50. Hienerwadel, R.; Grzybek, S.; Fogel, C.; Kreutz, W.; Okamura, M.Y.; Paddock, M.L.; Breton, J.; Nabedryk, E.; Mäntele, W. *Biochemistry* **1995**, *34*, 2832-2843.
51. Tiede, D.M.; Vasquez, J.; Cordova, J.; Marone, P.A. *Biochemistry* **1996**, *35*, 10763-10775.
52. Li, J.; Gilroy, D.; Tiede, D.M.; Gunner, M.R. *Biochemistry* **1998**, *37*, 2818-2829.
53. Baciou, L.; Sinning, I.; Sebban, P. *Biochemistry* **1991**, *30*, 9110-9116.
54. Nabedryk, E.; Breton, J.; Hienerwadel, R.; Fogel, C.; Mäntele, W.; Paddock, M.L.; Okamura, M.Y. *Biochemistry* **1995**, *34*, 14722-14732.
55. Jaques Breton, J.; Navedryk, E.; Mioskowski, C.; Boullais, C. In *The Reaction Center of Photosynthetic Bacteria -- Structure and Dynamics*; Michel-Beyerle, M.E., Ed.; Springer: Berlin, 1996; pp381-394.
56. Breton, J.; Navedryk, E. *Photosynthetic Research*, **1998**, *55*, 301-307.
57. Klapper, I.; Fine, R.; Sharp, K.A.; Honig, B. *Proteins: Struct., Funct., Genet.* **1986**, *1*, 47-59.
58. Sitkoff, D.; Sharp, K.A.; Honig, B. *J. Phys. Chem.* **1994**, *98*, 1978-1988.
59. Hagler, A.T.; Huler, E.; Lifson, S. *J. Am. Chem. Soc.* **1973**, *96*, 5319-5327.
60. Yang, A.S.; Gunner, M.R.; Sampogna, R.; Sharp, K.; Honig, B. *Proteins* **1993**, *15*, 252-265.
61. Kraulis, P.J. *J. Appl. Cryst.* **1991**, *24*, 946-950.

Chapter 6

Modeling the First Electron Transfer from Q_A to Q_B in Reaction Center Proteins from *Rb. sphaeroides*

E. G. Alexov and M. R. Gunner*

Department of Physics, City College of New York, 138th Street and Convent Avenue, New York, NY 10031

The first electron transfer in reaction centers from *Rb. sphaeroides* was modeled using the Multi Conformation Continuum Electrostatics (MCCE) method. It was found that the calculated free energy and proton binding of the electron transfer from Q_A^- to Q_B compare well with experiment from pH 5 to 11. At pH 7 the free energy (ΔG_{AB}) is measured as -65 meV and calculated to be -80meV. It was found that a cluster of acids (GluL212, AspL210, and L213) and SerL223 near Q_B play important roles. A simplified view shows this cluster has a single negative charge (on L213 which hydrogen bonds to Ser) in the ground state. In the Q_B^- state the cluster still has one negative charge, now on L210, further from Q_B. AspL213 and SerL223 move so L223 can hydrogen bond to Q_B^-. These rearrangements plus other changes throughout the protein make the reaction energetically favorable. The importance of GluL212 and AspL213 for the first electron transfer was tested by modeling several mutants in positions L212 and L213. It was found that substitution of GluL212 with Ala has a smaller effect than the substitution at AspL213. GluL212 is calculated to be protonated in both the $Q_AQ_B^-$ and $Q_A^-Q_B$ states, while AspL213 is ionized in the $Q_A^-Q_B$ state. Thus, mutation of AspL213 changes the ΔG_{AB} of the reaction predominately by changing of the free energy of $Q_A^-Q_B$ state.

© 2004 American Chemical Society

Introduction

The reactions of photosynthesis are initiated by the absorption of a photon which starts a series of electron transfers (1-4). The electron transfer from P* to H_L forming $P^+H_L^-$ and the subsequent electron transfer to the primary quinone to form $P^+H_LQ_A^-$ takes place with life-times of 2ps and 120 ps respectively. Such processes are fast enough that the protein can respond little to the electron's motion and therefore they can be considered as events that occur in a fairly rigid protein. Thus, they are sensitive to the preexisting protein arrangement (5) and protein electrostatic field (6). In contrast, the electron transfer from Q_A^- to Q_B occurs on the microsecond time scale, which allows the protein to rearrange itself in response to the new charge distribution(7). The challenge is to identify these motions and to evaluate their importance for the protein function.

In Rb. sphaeroides RCs both Q_A and Q_B are ubiquinone-10. However, despite their chemical identity, their function is different. Electron transfer from the ubiquinone in the Q_A site to the identical cofactor in the Q_B site is favorable, with a free energy of -65 meV (-1.5kcal/mol) (7). Thus, the differences between Q_A and Q_B must be controlled by their surroundings. Studies of site-directed mutations have identified a number of residues as being functionally important for electron and proton transfers to Q_B (reviewed by Okamura and Feher (8) and Takahashi and Wraight (3)). A number of mutations change the free energy of the electron transfer while others change proton uptake rates. Close to physiological rates and reaction energetics have been recovered by second-site revertants (9, 10). The ability of single residues to modify the reaction shows that in wild-type RCs a few residues are crucial. However, the ability of revertants to recover function shows that the pathway for proton uptake and the distribution of surrounding charged amino acids, which tune the correct reaction free energy is not unique.

The functional properties of RCs with changes at the site of two acidic residues, AspL213 and GluL212, have been extensively investigated in Rcs from Rb. capsulatus (11) and Rb. sphaeroides (12, 13). In both species GluL212 has little impact on the first electron transfer at physiological pHs, while AspL213 plays a crucial role. In Rb. sphaeroides, AspL213 (situated 5Å from Q_B) was shown to facilitate the delivery of the first proton to Q_B (14, 15). Its removal also changes the free energy gap between Q_A^- and Q_B^- states stabilizing the product (ΔG_{AB} = –110meV), by about 40meV from that found in WT RCs (ΔG_{AB} = -70meV) (14, 15). The removal of AspL213 also changes the pH dependence of the free energy. In native RCs the reaction is pH independent (pH 6-8) and becomes less favorable at higher and more favorable at lower pH. In RCs lacking AspL213, the free energy is significantly less pH dependent. Replacing GluL212 or AspL213 with other residues yields photosynthetically-incompetent strains (1). Some photocompetent phenotypic revertants derived from strains lacking AspL213 and GluL212 carry compensating mutations that are situated 10 to 15 Å away from Q_B (9).

The recent analysis of wild type (WT) and mutant *Rb. sphaeroides* RCs with the MCCE (Multi Conformation Continuum Electrostatics method (*16, 17*) provided a good match with experimental data for the ΔG_{AB} and protein uptake on the formation of Q_A^- and Q_B^-. With the exception of the calculations by Rabenstein et al. (*18*) all other numerical calculations have failed to reproduce the favorable free energy of electron transfer from Q_A^- to Q_B (*19-21*). The MCCE method combines calculation of residue ionization states and conformation as a function of the cofactor ionization state and the pH (*16, 17, 22*). A cluster of strongly interacting residues (L210Asp, GluL212, AspL213 and SerL223) was found to play an important role in the reaction (*19-21, 23*).

Studies of mutations focus on the role of specific residues. However, the positions and ionization states of all acidic and basic amino acids, micro-dipole orientation of polar residues, and the motions of the protein in response to changes in charge all affect the relative free energy of the different quinone reduction states. Recently, several mutants were explicitly modeled with the MCCE method (*17*). Substitution of residues at sites close to Q_B (GluL212, AspL213) as well as at more distant sites (ArgL233 and AsnM44) that were identified in phenotypic revertants (*10, 24*) were considered. The calculations found that change in charge of residues near the mutation site preserved the same net charge in mutants as in WT RCs.

Methods

The Multi Conformation Continuum Electrostatics (MCCE) method.

The atomic structure of the reaction centers from *Rb. sphaeroides* 1AIJ (*25*) (first molecule in the unit cell) from the Brookhaven Protein Data Bank (*26*) provided the framework coordinates of all backbone and non-polar side chain atoms for the calculations. The MCCE added pre-assigned alternate positions and charges to the side chains, cofactors, and buried water molecules. Each of these choices for a given residue charge and position is called a conformer. Acidic and basic residues have ionized and neutral conformers. Neutral acids, hydroxyl residues and waters have conformers with different proton positions. Buried waters can have their oxygens in alternate positions within a binding site. Each microstate of the protein has one conformer for each residue, cofactor, and water. Monte Carlo sampling calculates the probability of realizing each of initially suggested conformers in a Boltzmann distribution of microstates. The strength of the MCCE method is that atomic positions and ionization states are allowed to come to equilibrium in a single, self-consistent calculation (16,17).

The conformer occupancy was determined under a number of different conditions. The redox state of the protein was fixed for different simulations. Thus, the ground state calculations required that all microstates have only neutral P, Q_A, and Q_B. Separate Monte Carlo calculations were carried out for 3 redox states (ground, Q_A^- and Q_B^-) at 7 different pHs (from 3.5 to 11). Also in $P(Q_A$ or $Q_B)^-$ calculations all microstates have either Q_A^- or Q_B^- but never have both quinones reduced. The fraction of the protein with Q_A^- or Q_B^- at equilibrium can then be determined within these calculations.

Mutations were introduced in the structure by generating additional side chains at the appropriate positions on the WT backbone (17). Coordinates of the mutated amino acids were built using the Turbo-Frodo program (27). Thus, in addition to the GluL212 side chain coordinates, the input file contains additional atom coordinates for GluL212→Gln and GluL212→Ala. One extra conformer at position L213 for the AspL213→Ala mutant was introduced.

Results

The proton uptake and the free energy of electron transfer.

The free energy of Q_A^- to Q_B electron transfer (ΔG_{AB}) and the proton uptake caused by this reaction are coupled and they will be discussed together. Figure 1a shows the experimental free energy (28, 29) pH dependence compared with that was calculated using the equilibrium constant in the $P(Q_A^-$ or $Q_B^-)$ calculations as described in Alexov & Gunner (16). There is good agreement between experiment and calculations over the entire pH range. At pH 7 the measured ΔG_{AB} is -65meV (7) while the calculated value is -80±6 meV. The reaction becomes more favorable below pH 6 and less favorable at pH 9 and above in both experiment and calculation.

The one major feature of the reaction is the pH independence of the free energy at physiological pHs seen in both experiment and numerical calculations. If the free energy is pH independent then the proton uptake in the same pH range should be practically zero(30). Figure 1b, shows the calculated and measured proton uptake upon the first electron transfer from Q_A^- to Q_B. At physiological pHs the proton uptake is close to zero while at low and high pH the proton uptake is significant.

Figure 1b shows also the contribution to the proton uptake of the acidic cluster near Q_B. The cluster is composed of GluL212, AspL213 and L210 which are surrounded by many other acids and bases. GluL212:AspL213 have one charge and one proton in the ground state while both acids are protonated in the Q_B^- state. So over the entire pH range studied here, a proton must be bound by L212:L213 on Q_B reduction. Between pHs 7 to 9 this is accomplished by an intra-cluster proton shift as most of this proton is donated by AspL210. Figure 1b describes the proton uptake on electron transfer from Q_A^- to Q_B comparing the role of AspL210,

AspL213, the rest of the protein and the solvent. At pH 8 L213 binds 0.95 H$^+$/RC and L210 releases 0.75 H$^+$/RC. Of the extra 0.20 H$^+$/RC bound by L213, 0.15 is transferred from other sites in the protein each contributing <0.03 H$^+$/RC, while only 0.05 is donated from solution. The proton shift from residues surrounding the cluster adds to the buffer capacity ensuring that the net proton uptake into the protein is smaller than the increase in the cluster's protonation. Proton shifts from surrounding residues play a larger role at higher and lower pH's. Below pH 6, L210 holds onto its proton in the Q_B^- state so protonation of L213 requires transfer from the protein and the solvent. Above pH 9 L210 is largely ionized in both ground and Q_B^- states so again proton import into the cluster must occur.

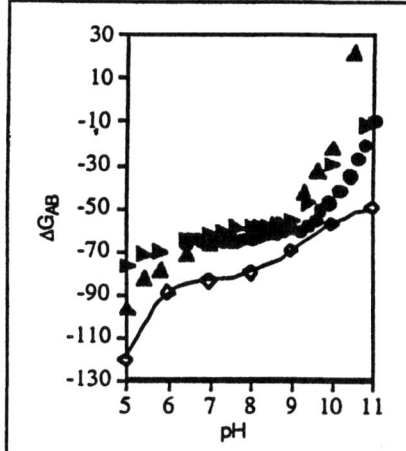

Figure 1a. The calculated free energy of Q_A^- to Q_B electron transfer. Experimental data taken from Ref. (10,28,36) represented by the symbols.

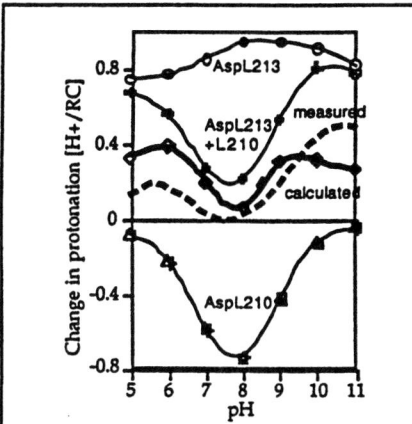

Figure 1b. Proton shift from AspL210 to L213 upon Q_A^- to Q_B electron transfer are shown. Experimental data is shown as broken curve and all other curves are labeled

Changes in the protein between the $Q_A^- Q_B$ and $Q_A Q_B^-$ states.

One conclusion of the calculations is that the reaction ΔG_{AB} would be quite unfavorable without changes in ionization and conformation (Figure 2a,b). Figure 2a represents the distribution of the of changes of conformation occupancy and 2b shows their spatial distribution. Residues that undergo large changes in conformer distribution are located in and around Q_B and in several nearby water channels.

There is a shift in a proton from L210 to L213. In addition there is reorientation of SerL223, and AsnM44. The occupancy of the site of water 6 which lies between AspL213 and L210 decreases by 56% and there is a loss in occupancy of the 4 water sites in the channel leading to the surface which ends in water 281 (numbering from 1AIJ). Simultaneously, with these large scale changes, many residues and waters shift their conformation distribution by a small amount. These small changes can be seen as points just off the diagonal in Figure 2a and as small balls in Figure 2b.. The interesting observation is although both Q_A and Q_B change charge in this reaction, most of the changes are situated within the Q_B pocket. This highlights the protein inhomogenety which is necessary for the RCs function. The Q_A pocket is

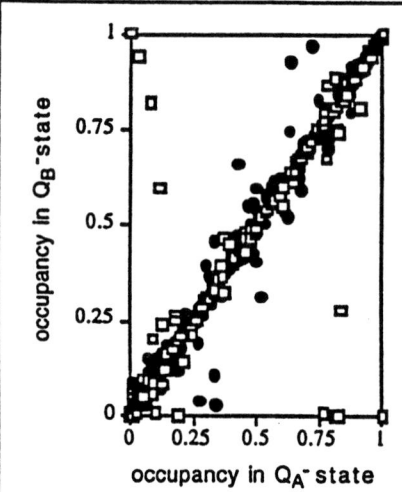

Figure 2a Conformer occupancy in Q_A^- state vs. occupancy in Q_B^- state at pH=7. Squares: protein; dots: water conformers. The points on the diagonal do not undergo conformation changes: points on the right of diagonal lower and points on the left increase conformer occupancy upon electron transfer.

Figure 2b Spatial distribution of the conformer occupancy changes upon the first electron transfer from Q_A^- to Q_B^- at pH=7. All waters are shown as dots. Each residue or water changes conformer distribution is shown as one ball; the size of the ball shows the magnitude of the change (<5%, 5-20% and > 20%).

much more rigid, while the Q_B pocket is flexible and capable of adopting a variety of ionization and conformation states. The largest changes are protonation of AspL213 (large; dark gray ball) and ionization of AspL210 (large; light gray). The hydroxyl reorientation of SerL223 and AsnM44 dipole reorientation (large; gray) are also seen to the left of the quinone. Waters within several water channels parallel to the membrane surface which undergo conformation changes are seen (*31*). Waters within the water channel perpendicular to the surface also show conformation changes.

Role of the cluster near Q_B.

A cluster of several groups within Q_B pocket was identified as playing a crucial role on the electron transfer from Q_A^- to Q_B. The cluster includes GluL212, AspL213, AspL210 and serL223. In the numerical calculations GluL212 changes charge only at very high pHs (> 10) and therefore will not be considered separately but instead as an acid closely coupled to AspL213.

The behavior of the cluster differs at physiological pHs and at low and high pHs.

In the physiological pH range and in the $Q_A^-Q_B$ state, AspL213 is practically fully ionized, while AspL210 is mostly protonated. When the electron is transferred from Q_A^- to Q_B at pH 7 to 8 there is significant rearrangement of both the atomic positions and charges in this cluster (Figure 3a). Asp L213 binds a proton, breaking the hydrogen bond to SerL223 which now makes a hydrogen bond to Q_B^-. Asp L213 and L212 have one proton in the ground state or in the Q_A^- state. However, in the Q_B^- state these 2 acids are fully protonated (Figure 3a). Thus Q_B plus L212, and L213 can only support one negative charge. However, when L210 is added than at physiological pHs this cluster of four groups can sustain a charge of -1 or -2. Thus, on electron transfer from Q_A^- the cluster goes from a charge of -1 with $[Q_B^0 + (L212:L213)^{-1} + L210^0]$ to a charge of -2 with $[Q_B^{-1} + (L212:L213)^0 + L210^{-1}]$.

At low pH (pH=3.5) the scenario is different. While the ionization and conformation states in the $Q_A^-Q_B$ state are the same at pH 3.5 and 7, the changes induced by the electron transfer are different. As at physiological pHs, AspL213 becomes protonated in the $Q_AQ_B^-$ state, but at low pH this does not involve ionization of AspL210, because the pK_a of AspL210 in the $Q_AQ_B^-$ state is bigger than 3.5. Proton transfer to AspL213 occurs from the surroundings most prominently from AspM17 (0.15H$^+$). Thus, AspL213 must now binds protons from the solution, as a result the free energy becomes pH dependant.

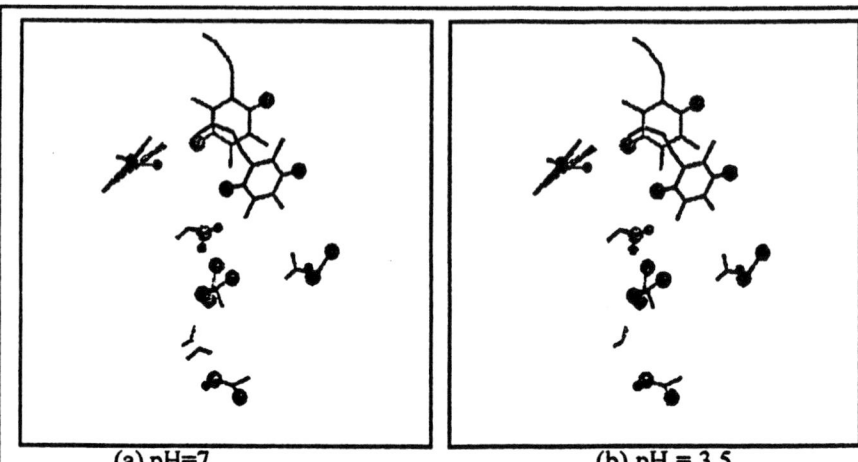

Figure 3. The most occupied conformers of several residues within the Q_B pocket that undergo either conformation or ionization changes on the electron transfer. The most occupied conformers in Q_A^- state are light gray and those in Q_B^- state are dark gray. Residues shown (moving downward on the left): TyrL222, SerL223, (AspL213 and GluL212), water 6 and AspL210. Q_B distal (on the top) and proximal (lower) positions.

Mutations at L212 and L213.

The role of GluL212 and AspL213 in determining ΔG_{AB} was tested by modeling two mutants (*17*). Figure 4 shows the pH titration of the free energy of the electron transfer from Q_A^- to Q_B. Figure 4a shows the ΔG_{AB} in the GluL212→Gln mutant as a function of pH. There is good agreement between the calculated and the experimental ΔG_{AB} values (*11, 13, 32*). The experimental observation that ΔG_{AB} for the WT and GluL212→Gln RCs differ only slightly at pH 7 led to the suggestion that GluL212 is protonated at this pH in the ground, Q_A^- and Q_B^- states (*32*). This is consistent with the MCCE calculations of WT RCs (*16*). The small difference in the ΔG_{AB} values in the mutant results from replacing the uncharged protonated Glu^0 by Gln. Detailed analysis of the protein ionization and conformation distribution in the mutant and WT RCs suggest almost identical pair-wise interaction energies of the dipoles of Gln and Glu^0 with Q_A^- and Q_B^-, as well as with individual residues in the protein.

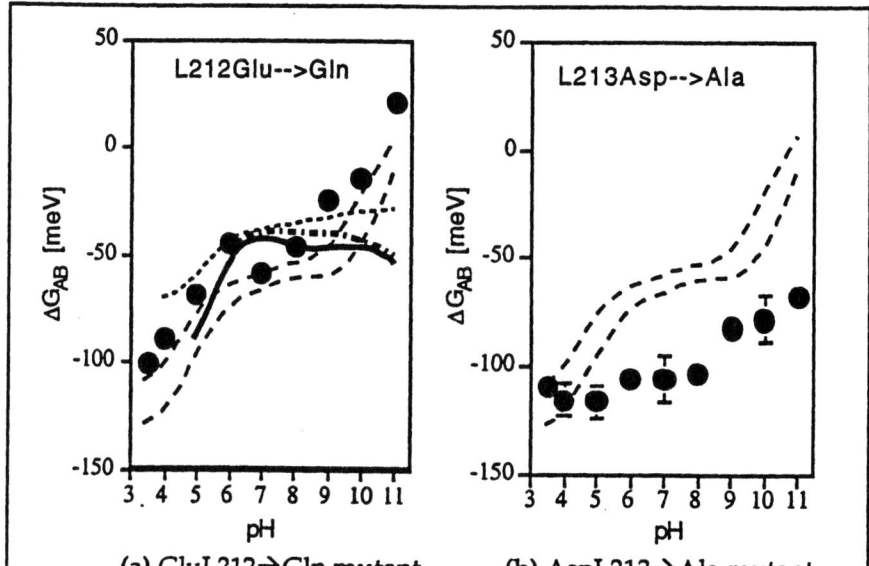

(a) GluL212→Gln mutant   (b) AspL213→Ala mutant

Figure 4.. Calculated free energies are shown by the filled circles with associated error bars. The envelope of the experimental data from WT RCs from Figure 1a is given by the two dashed lines. The solid line represents the experimental data from (35) (*Rb. caplulatus* RCs), dotted line from (30) and dash-dotted line from (34) (*Rb sphaeroides* RCs).

Above pH 9 in the mutant, the experimental ΔG_{AB} remains pH independent while in the calculations the mutant behaves like WT RCs. In the WT RCs GluL212 remains neutral above pH 11. It may be that partial ionization of GluL212 in WT RCs at a pH above 9.5, missed in the calculations, accounts for the experimentally observed difference at high pH between the WT and the GluL212→Gln mutant.

The AspL213→Ala single mutant has not been constructed, therefore no experimental data are available, but the results can be compared qualitatively with the AspL213→Asn mutant (*14, 15*). Here ΔG_{AB} is found to be -50meV more favorable than in the WT RCs. At pH 7, in the AspL213→Ala RCs, the calculated ΔG_{AB} is -25 meV smaller than that calculated for WT RCs and it is pH independent below pH 9. Above pH 9, ΔG_{AB} becomes less favorable.

Two major differences in ionization states were found comparing WT and the mutant RCs in the ground state at physiological pHs. AspL213, which is fully ionized in the WT RCs, is removed in the mutant. Its negative charge is then transferred to L210Asp (75%) and to GluL212 (15%), which are 25% ionized (L210Asp) and fully (GluL212) neutral in WT RCs. Thus in the mutant, L210Asp is fully ionized and GluL212 is partially ionized in the ground state at pH 7. The net charge of the cluster of the three residues is -1.25 in WT and -1.15 in mutant RCs.

The electron transfer from Q_A^- to Q_B is calculated to be more favorable in the mutant than in WT RCs. Residue ionization and conformation in the Q_B^- state are little changed on removing Asp which is neutral when Q_B is reduced. In contrast, in the Q_A^- state AspL213 is fully ionized. Thus, removing AspL213 changes the ionization of surrounding residues in the ground and Q_A^- but not the Q_B^- states. Thus, the calculated ΔG_{AB} notably increases mostly by destabilizing the reactant $Q_A^-Q_B$ state, rather than by stabilizeng the product, $Q_AQ_B^-$ state in the the AspL213→Ala mutant. In the $Q_A^-Q_B$ state, the carboxyl group of AspL213$^-$ makes a strong hydrogen bond to the hydroxyl of L223Ser (33), which is lost in the mutant. The loss of the hydrogen bond destabilizes the $Q_A^-Q_B$ state in the mutant. In the $Q_AQ_B^-$ state the L223Ser makes a hydrogen bond to Q_B^- which is not affected by the removal of AspL213.

DISCUSSION

The electron transfer from Q_A^- to Q_B in bacterial RCs has been the subject of earlier theoretical studies considering both *Rb. sphaeroides* and *Rps. viridis*. RCs. All previous calculations in RCs either did not report the free energy of electron transfer from Q_A^- to Q_B (ΔG_{AB}) (*20, 21*) or provided a free energy that was very unfavorable (*19*). The only one exception is the calculations of Rabenstein et al. (*18*). All these earlier numerical simulations were carried out using standard methodology for pK_a calculations that presumes a rigid protein structure. Waters were removed, hydroxyl protons were minimized in the protein ground state, and only ionization states of acidic and basic residues change when the redox state of the protein changes. The rigid protein calculations of Rabenstein et al. do reproduce the measured ΔG_{AB} at the pH of their calculations. Several specific changes in the structure were made. Q_B was moved and the side chain of GluL212 was rotated. No pH dependence of the free energy was reported. Also newly calculated charges were used for the quinones. In comparison to the calculations reported here movement of GluL212 would be expected to produce the most significant effect by changing the energies of different charge states of the active cluster.

Each calculation has found a cluster of acids in the Q_B site that is not fully ionized. The redistribution of protons within the cluster and uptake of protons into the cluster from the rest of the protein or solvent occurs when Q_B is reduced. The residues GluL212 and AspL213 and L210 interact strongly with each other and with Q_B in *Rb. sphaeroides* RCs. These residues have also been recognized as important by site directed mutagenesis studies (*12-15, 34, 35*). In *Rps. viridis* L213 is an Asn not an Asp so the groups that make up the cluster must be different. The Glutamic acids L212, H177 (H173 in *Rb. sphaeroides*), and M234 are now identified as the active cluster (*20*). Thus, there is a consensus about the presence and identity of the residues in the clusters in these RCs. In the ground state at physiological pH each calculation finds a net charge of -1 (here) to -2 distributed amongst the 3 acids. This provides 1 to 2 protons bound in the cluster that can change position when Q_B is reduced, diminishing the requirement for proton uptake from solution to stabilize the quinone charge.

At physiological pHs, Q_A reduction causes several small ionization changes of acidic groups in the Q_B pocket, which are more than 15Å away. Thus, the region of the protein near Q_A is rather rigid and it does not undergo large ionization or conformation changes. There are instead small changes dispersed over much of the protein with no conformer probability changing by as much as 10%.

In contrast Q_B reduction causes large changes in ionization states, as well as changes in hydroxyl, side chain and water conformation which are localized in the protein near Q_B. Thus, Q_B looks like an environment with a relatively high effective dielectric constant. Local rearrangements screen more distant sites from the effects of a new charge, and these rearrangements stabilize the charge so no additional proton uptake is required.

The calculations duplicate the experimentally determined pH dependence of ΔG_{AB}. In the physiological pH range the acid cluster of L210, L212, and L213 plus SerL223 play a pivotal role. GluL212 and AspL213 are practically symmetrically situated around Q_B and so interact with the quinone by almost the same amount. However, AspL210 interacts with AspL213 more strongly than with GluL212. Thus, AspL213 can control the ionization of AspL210 while GluL212 is less effective. In the ground state and the Q_A^- state there is one proton on L212 and L213. If the ionized acid is L213 then L210 is also protonated. When Q_B is reduced L212 and L213 must both be protonated. When possible the proton comes in from L210, 10Å away from Q_B. However, if L210 is ionized in the Q_A^- state L213 is protonated by transfer from the rest of the protein and solution. L210 is ionized in the Q_A^- state at higher pH, and so significant proton uptake is seen. In addition, L210 will be ionized if L213 is neutral in response to L212 ionization. Solution proton uptake will now be more strongly coupled to Q_B reduction and ΔG_{AB} will become pH dependent. This may account for the larger pH dependence of ΔG_{AB}

found in earlier calculations where L212 and L210 are partially ionized in the ground state.

The roles of the residues in the cluster are in agreement with the picture obtained by site-directed mutagenesis experiments which suggest that AspL213 and SerL223 are involved in the first proton transfer to Q_B (*14, 15*). The calculations also find this chain of residues moving into a configuration that can deliver the proton to Q_B^-. Here a proton is transferred from protonated AspL210 to AspL213 when Q_B^- is formed. This proton transfer breaks the hydrogen bond between AspL213$^-$ and SerL223 reorienting the Ser hydroxyl hydrogen so that it makes a hydrogen bond to Q_B^-(*33*). L223 is then properly situated and L213 is protonated and prepared for proton transfer to Q_B^-. Water 6 (1AIJ (*25*) water numbering) connects L210 and L213. On Q_B reduction the water occupancy of this site is reduced. This change propagates through a channel, parallel to the membrane, with 4 water sites ending at water 281. The second channel, perpendicular to the membrane, starts with water 325, near AspL213 and leads to the surface. Members of this channel also change conformation upon quinone reduction. Thus, the calculations show at least two proton pathways that can deliver protons to the active cluster. In addition, there is another channel parallel to the membrane that connects water 6, via the side chain of ArgL217, to water 42 and then to the surface.

Each of the mutations discussed here deletes an ionizable group from the protein. The formal pair-wise interactions of these residues, if charged, with Q_B^- are 330meV for GluL212 and 300meV for AspL213. However measured and calculated $\Delta\Delta G_{AB}$ in the mutants vary by less than ±60meV. Two mechanisms play a role in reducing the $\Delta\Delta G_{AB}$ (difference between ΔG_{AB} calculated in mutant and in native RCs). First, both residues subject to mutation are neutral in the $Q_A Q_B^-$ state. So, these large pair-wise interactions with Q_B^- never contribute to ΔG_{AB}. In addition, when the group is ionized in the $Q_A^- Q_B$ state (AspL213), its deletion causes changes in the ionization within a particular cluster of strongly interacting groups. Ionization changes within the cluster keep the net charge unchanged, reducing the effect of the mutation. Each new protein charge distribution does change ΔG_{AB} compared to WT RCs by a modest, but easily measurable amount. However the change would be much larger without protein rearrangement that maintains the net charge of the protein.

In each mutant the change in ΔG_{AB} can be caused by changes in the free energy levels of either of the semiquinone states. The effect of a given mutation depends on the initial role that the native side chain plays in determining the free energy levels of the $Q_A^- Q_B$ and $Q_A Q_B^-$ states rather than if its position is nearer Q_A or Q_B. Thus, several mutations have the largest effect on the $Q_A^- Q_B$ state despite the residue being close to Q_B. For example, initial analysis of AspL213→Ala mutant proposed that the more favorable ΔG_{AB} resulted from removing a charge 4Å away from Q_B. However, AspL213 doesn't contribute significantly to the energy of the $Q_A Q_B^-$ state, because it is fully protonated in this state. Rather, it makes significant contributions to the free energy of the Q_B pocket in the $Q_A^- Q_B$ state (17).

ACKNOWLEDGMENT

We would like to thank P. Sebban, L. Baciou, D. Hanson and M. Schiffer for providing us with experimental data on several mutants and for their active contribution to the studies of mutants. We would like thank George Feher and Ed Abresh for access to the 1AIJ and 1AIG structures prior to their deposition in the protein data bank, Jacques Breton and Eliane Nabedryk, Colin Wraight, Mel Okamura and Mark Paddock for helpful discussions. We are grateful for the support of NSF MCB 9629047 and NATO LST.CLG 975754.

References

1. Blankenship, R. E., Madigan, M. T., and Bauer, C. E. *Anoxygenic Photosynthetic Bacteria*, (1995), Vol. 2, Kluwer Academic Publishers.
2. Okamura, M. Y., and Feher, G. in *Anoxygenic Photosynthetic Bacteria* (Blankenship, R., Madigan, M., and Bauer, C., Eds.) (1995), pp 577-593, Kluwer Academic Publishers, Dordrecht.
3. Takahashi, E., and Wraight, C. A. *Advances in Molecular and Cell Biology* (1994), *10*, 197-251.
4. Sebban, P., Maroti, P., and Hanson, D. K. *Biochimie* (1995), 77, 677-694.
5. Hanson, L. K., Thompson, M. A., Zerner, M. C., and Fajer, J. *Theoretical models of electrochromic and environmental effects on bacterio-chlorophylls and -pheophytins in reaction centers*, (1988), Plenum Publishing Corporation.
6. Gunner, M. R., Nicholls, A., and Honig, B. *J. Phys. Chem.* (1996), *100*, 4277-4291.
7. Graige, M. S., Paddock, M. L., Feher, G., and Okamura, M. Y. *Biochemistry* (1999), *38*, 11465-11473.
8. Okamura, M. Y., and Feher, G. (1992) *Annu. Rev. Biochem.* 61, 861-896.
9. Sebban, P., Maroti, P., Schiffer, M., and Hanson, D. K. *Biochemistry* (1995), *34*, 8390-8397.
10. Hanson, D. K., Baciou, L., Tiede, D. M., Nance, M., Schiffer, M., and Sebban, P. *Biochim. Biophys. Acta* (1992), *1102*, 260-265.
11. Maroti, P., Hanson, D. K., Baciou, L., Schiffer, M., and Sebban, P. *Proc. Natl. Acac. Sci.* (1994), *91*, 5617-5621.
12. Paddock, M. L., Feher, G., and Okamura, M. Y. *Biochemistry* (1997), *36*, 14238-14249.
13. Takahashi, E., and Wraight, C. A. *Biochemistry* (1992), *31*, 855-866.

14. Takahashi, E., and Wraight, C. A. *Biochim. Biophys. Acta* (1990), *1020*, 107-111.
15. Paddock, M. L., Rongey, S. H., McPherson, P. H., Juth, A., Feher, G., and Okamura, M. Y. *Biochemistry* (1994), *33*, 734-745.
16. Alexov, E., and Gunner, M. *Biochemistry* (1999), *38*, 8253-8270.
17. Alexov, E., Miksovska, J., Baciou, L., Schifer, M., Hanson, D., Sebban, P., and Gunner, M. *Biochemistry* (2000), *submitted*.
18. Rabenstein, B., Ullmann, G. M., and Knapp, E.-W. *Eur. Biophys. J.* (1998), *27*, 626-637.
19. Beroza, P., Fredkin, D. R., Okamura, M. Y., and Feher, R. *Biophys. J.* (1995), *68*, 2233-2250.
20. Lancaster, C. R. D., Michel, H., Honig, B., and Gunner, M. R. *Biophys. J.* (1996), *70*, 2469-2492.
21. Gunner, M. R., and Honig, B in *The Photosynthetic Bacterial Reaction Center: Structure, Spectroscopy and Dynamics II* (Breton, J., and Vermeglio, A., Eds.) . (1992), pp 403-410, Plenum, New York.
22. Alexov, E. G., and Gunner, M. R. *Biophys. J.* (1997), *72*, 2075-2093.
23. Grafton, A. K., and Wheeler, R. A. *J. Phys. Chem* (1999),*103*, 5380-5387.
24. Maroti, P., Hanson, D. K., Schiffer, M., and Sebban, P. *Nature Struc. Biology* (1995), *2*, 1057-1059.
25. Stowell, M. H. B., McPhillips, T. M., Rees, D. C., Soltis, S. M., Abresch, E., and Feher, G. *Science* (1997), *276*, 812-816.
26. Bernstein, F. C., Koetzle, T. F., Williams, G. J. B., Meyer, E. F., Brice, M. D., Rodgers, J. R., Kennard, O., Shimanouchi, T. F., and Tasumi, M. *J. Mol. Biol.* (1977), *112*, 535-542.
27. Roussel, A., and Cambillan, C. *Turbo-Frodo* (1991), Vol. 86, Mountain View, CA.
28. Maroti, P., and Wraight, C. A. *Biochim. Biophys. Acta* (1988), *934*, 329-347.
29. McPherson, P. H., Okamura, M. Y., and Feher, G. *Biochim. Biophys. Acta* (1988), *934*, 348-368.
30. Schellman, J. A. *Biopolymers* (1975), *14*, 999-1018.
31. Sham, Y. Y., Muegge, I., and Warshel, A. *Proteins: Structure, Function, and Genetics* (1999), *36*, 484-500.
32. Paddock, M. L., Rongey, S. H., Feher, G., and Okamura, M. Y. *Proc. Natl. Acad. Sci. USA* (1989), *86*, 6602-6606.
33. Lancaster, R., and Michel, H. *Structure* (1997), *5*, 1339-1359.
34. Miksovska, J., Kalman, L., Schiffer, M., Maroti, P., Sebban, P., and Hanson, K. *Biochemistry* (1997), *36*, 12216-12226.
35. Hienerwadel, R., Grzybek, S., Fogel, C., Kreutz, W., Okamura, M. Y., Paddock, M. L., and Breton, J. *Biochemistry* (1995), *34*, 2832-2843.

Chapter 7

Dynamics of Electron Transfer Pathways in Redox Proteins

Ilya A. Balabin and José Nelson Onuchic

Department of Physics, University of California at San Diego, La Jolla, CA 92093-0319

Protein dynamics has been assumed to affect the rate of biological electron transfer reactions by controlling the Franck-Condon factor (Marcus theory). We show that the dynamics may also strongly affect the effective tunneling coupling for bridges with a destructive interference among the dominant electron transfer pathways. This new type of dynamical control is quantitatively described for two primary electron transfer reactions in a bacterial photosynthetic reaction center. We demonstrate that even small nuclear motions may lead to a tremendous increase in the reaction rate, thus providing high efficiency of redox proteins.

Electron transfer (ET) reactions are key steps in many bioenergetic events, particularly photosynthesis and respiration (1-3). Because of typically large (5 – 20 Å) tunneling distances, the effective tunneling coupling between the donor and the acceptor is mediated by the protein or cofactor matrix, and the reactions occur in the weak coupling limit. The reaction rates are therefore proportional to the Franck-Condon factor, which describes the probability for the donor and the acceptor to form a resonant activated complex, and the square of the effective coupling. While the nuclear factor has been reasonably well described by Marcus theory (4) and its semi-classical and quantum modifications, the effective coupling needs a substantially better theoretical understanding. First principle calculations are still very difficult even for medium size proteins, and when they are possible, it is often hard to identify what structural features are most relevant and how protein dynamics can affect the coupling. As so, simple models are still of major importance for investigating biological redox reactions, guiding detailed calculations and designing new redox proteins.

The standard theoretical tool utilized by experimentalists to estimate the matrix elements and to understand the tunneling mechanism in proteins has been the *Pathways* method developed by Beratan and Onuchic about ten years ago. Based on the central assumption that tunneling occurs via a dominant pathway tube inside the protein and that the decay through this tube can be quantified as a product of contributions from covalent bonds, hydrogen bonds and through space jumps(5-6), *Pathways* has been successful and had a large impact in the experimental community (1-3,7). However, *Pathways* predictions are limited by two major issues: the possibility of interference among multiple dominant paths is not included, and calculations are done for the protein crystallographic structure without considering how dynamics may affect tunneling. Although dynamical effects have been observed in some preliminary calculations that used both the *Pathways* and other methods (8-10), no consistent theoretical treatment has been developed yet.

We present a new quantitative approach to evaluate the *Pathways* limitations and to address the two questions raised above. Our approach includes identifying the dominant ET pathways, analyzing the interference among them and exploring the sensitiviy of the effective coupling to nuclear dynamics. Our results show that *Pathways* provides a reasonable estimate of the rate for ET reactions that are dominated by a single pathway tube or few tubes that interfere constructively. For proteins with a destructive interference among the dominant paths, the effective coupling is extremely sensitive to structural details and therefore to protein dynamics. In this case, tunneling may be controlled by protein conformations "far" from crystallographic, in which one of the tubes dominates the coupling, minimizing effects of destructive interference. We demonstrate this concept for two primary ET reactions in a bacterial photosynthetic reaction center (RC) (11-18): from the pheophytin (*Bph*) to the primary quinone (Q_A) and from the latter to the secondary quinone (Q_B). We provide a quantitative description of the dynamical effects, and we show that they may be critical for ET reaction efficiency.

Methods

Effective Tunneling Coupling Calculations

The ET reaction rate in the weak coupling limit is described by the Fermi Golden rule (4)

$$k_{ET} = \frac{2\pi}{\hbar} \langle T_{DA}^2 \rangle \, (F.C.), \qquad (1)$$

where T_{DA} is the effective tunneling coupling between the donor and acceptor, and $(F.C.)$ is the Franck-Condon density of states that describes the nuclear control of the energy gap. For simplicity, we will use the maximum reaction rate for comparison with the *Pathways* predictions and experiments, assuming that the nuclear factor is optimized.

For a given protein conformation, the effective coupling is (19-20)

$$T_{DA} = \sum_{i,j}^{bridge} (E_{tun} \, S_{Di} - H_{Di}) \, \tilde{G}_{ij} \, (E_{tun} \, S_{jA} - H_{jA}), \qquad (2)$$

where **S** and **H** are the electronic orbital overlap and Hamiltonian matrices, respectively, E_{tun} is the tunneling energy, indices i and j refer to bridge orbitals, and the transformed Green's function of the bridge $\tilde{\mathbf{G}} = (E_{tun} \, \mathbf{S} - \mathbf{H})^{-1}$ incorporates the effects of the nonorthogonality of the orbital basis (21). Modulating the orbital overlaps and couplings, protein dynamics makes T_{DA} change with time. As described by eq. (2), the dynamics of T_{DA} reflects two factors: the modulation of the overlaps and couplings between the donor (or acceptor) and the bridge orbitals, and the modulation of the Green's function. The former are dominated by strong short range interactions, which are robust to protein dynamics (22). The Green's function matrix elements, however, can be very sensitive to the nuclear motions (22), and they primarily determine the T_{DA} dynamics. As so, we can approximate the electronic term in eq. (1) as

$$\langle T_{DA}^2 \rangle \approx (E_{tun} \langle S_{Di_0} \rangle - \langle H_{Di_0} \rangle)^2 \langle (\tilde{G}_{i_0 j_0})^2 \rangle (E_{tun} \langle S_{j_0 A} \rangle - \langle H_{j_0 A} \rangle)^2. \qquad (3)$$

where i_0 and j_0 are the bridge orbitals connected to the donor and acceptor, respectively. The extent in which the nuclear dynamics affects the tunneling coupling can be quantitatively described by the *coherence parameter*

$$C = \frac{\langle \tilde{G}_{i_0 j_0} \rangle^2}{\langle \tilde{G}_{i_0 j_0}^2 \rangle}. \qquad (4)$$

Since $\langle \tilde{G}^2_{i_0 j_0} \rangle = \langle \tilde{G}_{i_0 j_0} \rangle^2 + \langle \delta \tilde{G}^2_{i_0 j_0} \rangle$, where $\langle \delta G^2_{i_0 j_0} \rangle$ is the dynamical mean square deviation, the coherence parameter is in the range between zero and one. In the limit that the dynamic variations of the Green's function are small, C approaches 1, while in the opposite limit, when the dynamical changes are large in comparison with the average values, C is close to zero. These two limits correspond to two opposite ET regimes, when the effective coupling is primarily controlled by the bridge structure and by the bridge dynamics, respectively.

Electronic Hamiltonian

The standard extended Huckel Hamiltonian (23) with a minimum basis set of Slater-type atomic orbitals were used in all calculations. The overlap and Hamiltonian matrices were computed for the RC fragment in the light conformation (12), which included Bph, Q_A, Q_B, the iron ion and the relevant protein environment: Trp^{L100}, Met^{M256}, Trp^{M252}, Met^{M218}, His^{M219}, His^{L190}, His^{L230}, Glu^{M234} and His^{M266}. The iron orbitals were modeled by p orbitals with the decay constants and orbital energies tuned up to reproduce the correct valence shell radii and ionization potentials, respectively (24). Since the iron ion was covalently bound to the six nearest coordination neighbors (12), this approximation only had a minor effect on the effective coupling between Q_A and Q_B. The diagonal matrix elements of the Hamiltonian were the ionization potentials of atomic orbitals $H_{ii} = \alpha_i$, and the off-diagonal elements were assigned as $H_{ij} = (K/2)(\alpha_i + \alpha_j) S_{ij}$, where $K = 1.75$ was the standard Huckel constant. The overlap and Hamiltonian matrices were converted to the bond orbital basis by using the standard transformations (25). The donor and acceptor orbitals were the ground state pheophytin and quinone ring π orbitals, i.e. normalized sums of all π orbitals orthogonal to the ring plane. The method of calculating T_{DA} described above can be used for an arbitrary Hamiltonian, and an extended Huckel level Hamiltonian is used in this article for simplicity.

Molecular Dynamics

To compute the average square effective coupling $\langle T^2_{DA} \rangle$ (eq. (3)), the Green's functions were calculated for a number of dynamical conformations ("snapshots"), which were obtained by running molecular dynamics (MD). The effective coupling was also computed for the crystallographic and the "average" conformations (nuclear coordinates averaged over the entire MD runs). The MD simulations were performed with the CVFF forcefield (26) for the RC fragment and all aminoacids within 8 Å from it. Being mostly interested in effects of small nuclear motions, we limited the dynamics by imposing a set of constraints and non-holonomic restraints. All backbone atoms were fixed, and all other heavy atoms were pulled to their positions in the crystallographic structure by harmonic restraining forces (tethered). The root mean square deviation (RMSD) of heavy atoms was controlled by choosing the tethering force constant. The RMSD between the crystallographic and the average conformation was about 0.35 Å, and the RMSD between any "snapshot" and the average conformation was about 0.25 Å. Notice that larger nuclear motions, which we do not allow here, may even change the tube structure, leading to even more pronounced dynamical effects. The RC fragment was first equilibrated for 3 ps, and then MD was run for 15 ps with a 0.5 fs time step. A "snapshot" conformation was saved every 50 ps, i.e. each MD run was described with a set of 301 snapshots.

Results and Discussion

The dominant pathways and the effective coupling for the ET reaction from Bph to Q_A are shown at Fig. 1. The energy range corresponds to the biologically relevant tunneling energies, and typical tunneling energies are in the middle of this range. The dominant pathways were identified by computing the effective coupling for the RC fragment (T_0) and a number of its modifications, when one aminoacid was removed in each modification (T_i for the i-th aminoacid removed). The relevance of i-th aminoacid was quantitatively described by the ratio $|(T_\circ - T_i)/T_\circ|$. Importance of individual orbital interactions that mediate through space jumps was characterized in a similar way. Since more than 80% of T_{DA}^2 is mediated by Trp^{M252}, the interference regime is defined by the through space jumps from Bph to the tryptophan and from the latter to Q_A (Fig. 1). The first jump is mediated by two interactions between the Bph and tryptophan σ orbitals, which form two parallel pathways with comparable strengths and the same phases, leading to a constructive interference. At the second jump, a strong interaction between the π orbitals on the tryptophan and Q_A carbons forms a single dominant pathway. A weaker alternative pathway (about 20% of T_{DA}^2) mediated by Met^{M218} and His^{M219} has a minor effect on the interference. The coupling is almost insensitive to details of the protein structure, indicating that the pathway interference is constructive.

Between Q_A and Q_B, the dominant pathways were identified in the same way as above (Fig. 2). The interference is highly destructive, since the paths between Q_A and His^{M219} are mediated by a hydrogen bond and a few strong through space jumps with similar strengths but different phases. The paths between His^{L190} and Q_B have a similar structure.

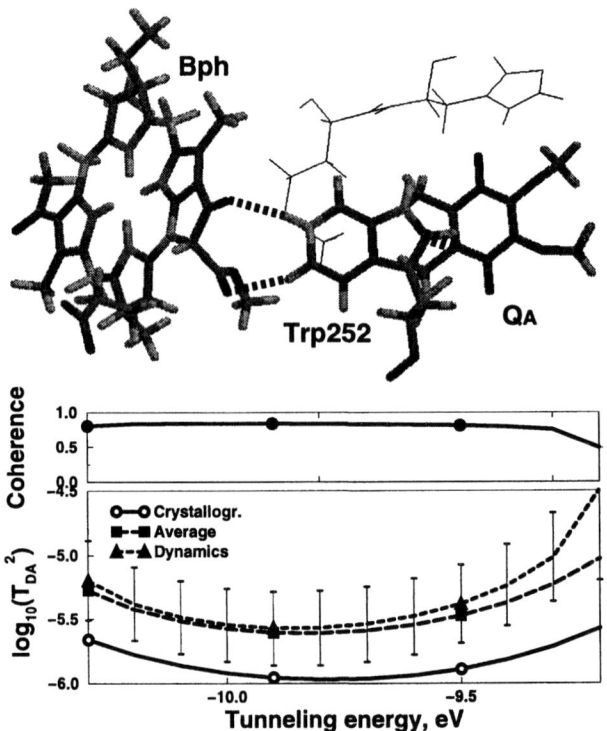

Figure 1: Electron transfer pathways between *Bph* and Q_A. Upper panel: the coherence parameter $C = \langle \tilde{G} \rangle^2 / \langle \tilde{G}^2 \rangle$ is close to 1, indicating that dynamical effects on the coupling are small. Lower panel: the effective couplings for the crystallographic and the average conformation and $\langle T_{DA}^2 \rangle$ are close. The coupling is controlled by the overall protein structure, and it is almost insensitive to structural details and nuclear dynamics.

In addition, the metal bonds with His^{L230}, Glu^{M234} and His^{M266} form a complex network of additional paths with different phases. These structural features together result in a pronounced destructive interference among the dominant pathways, which makes the coupling highly sensitive to small structural details.

These different interference regimes lead to different mechanisms that control the effective coupling for the two ET reactions. Because of a constructive pathway interference between Bph and Q_A, the coupling is almost insensitive to the structural details and therefore to dynamics. As shown in Fig. 1, the couplings for the crystallographic and the average conformation and $\langle T_{DA}^2 \rangle$ are very close. The *coherence parameter* (eq. (4)) is about 1, indicating that the dynamical effects are minor. The coupling is thus controlled by the overall bridge structure and not by structural details or nuclear dynamics. Conversely, the destructive interference between Q_A and Q_B makes the contributions from the dominant pathways almost cancel each other in the crystallographic or average conformation. Since the coupling is very sensitive to details of the structure, it is strongly modulated by the nuclear dynamics: the coherence parameter (eq. (4)) is close to zero (Fig. 2). Thus, the major contribution to the coupling is provided by conformations far from the "frozen", where we expect individual pathway tubes to dominate. The dynamics therefore leads to $\langle T_{DA}^2 \rangle$ almost three orders of magnitude larger than the square coupling for the crystallographic conformation (Fig. 2). This *dynamic amplification* only involves local nuclear motions, unlike the conformational gating, which requires a global motion of a whole structural group (12). Importantly, for the ET reactions with a destructive interference, one needs to perform the calculations with a number of snapshots: calculations based on a single conformation, independently of the method, may be grossly misleading.

The results above suggest that the *Pathways* method provides reasonable predictions for bridges with a constructive interference among the dominant pathways but may break down if the interference is destructive. To compare our results, *Pathways* calculations and experiments, we convert the effective couplings to the maximum rate (eq. (1)) using the optimized Franck-Condon factors: $k_{ET} \approx 5 \cdot 10^{16} \ (T_{DA}, eV)^2 \ s^{-1}$. We used the standard *Pathways* estimate of the effective coupling (5)

$$T_{DA} \approx (E_{tun} \ S_{Di} - H_{Di}) \ \tilde{G}_{ii} \prod_i \epsilon^C \prod_j \epsilon_j^{TS} \ (E_{tun} \ S_{jA} - H_{jA}), \quad (5)$$

where $\epsilon^C \approx 0.6$ is the decay per covalent bond, $\epsilon_j^{TS} \approx \epsilon^C exp(-\alpha(r_j - r_0))$ is the decay per TS jump, $\alpha \approx 1.7 \text{Å}^{-1}$ is the through space distance decay, r_j is the jump distance, and $r_0 \approx 1.4$ Å. The typical nearest neighbor orbital

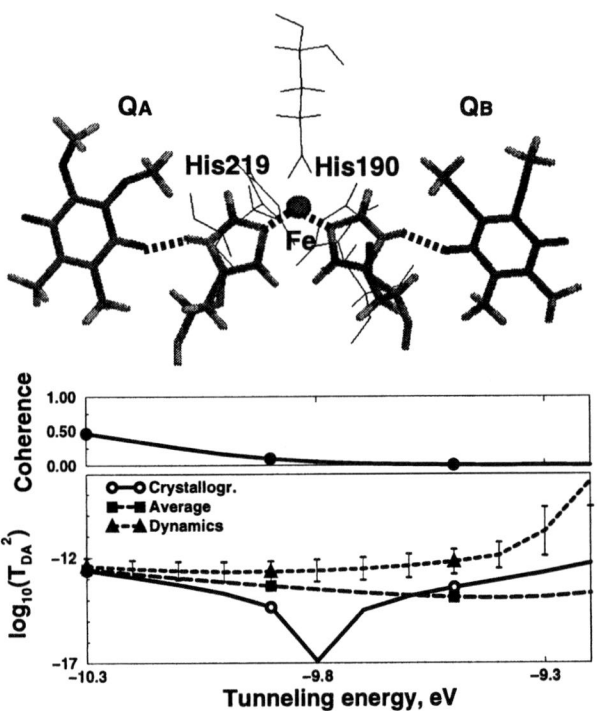

Figure 2: Electron transfer pathways between Q_A and Q_B. Upper panel: the coherence parameter (eq. (4)) is close to zero, indicating that the coupling is controlled dynamically. Lower panel: the effective coupling is sensitive to the structural details; the couplings for the crystallographic and the average conformations are substantially different. The dynamics increases the coupling by almost three orders of magnitude at the relevant energies.

overlaps and couplings used were $(E_{tun} S_{Di} - H_{Di}) \approx 5 \ eV$ (27), and the self Green's function matrix elements $\tilde{G}_{ii}$ were computed at the extended Huckel level. These parameters are in a good agreement with previous *Pathways* level calculations (7,28) and other approaches (29), providing the maximum rate about $10^{13} \ s^{-1}$ at distances about 3 Å.

Between Bph and Q_A, our calculations provide the maximum rate $10^{11} \ s^{-1}$, which is by about ten times larger than the experimental value $10^{10} \ s^{-1}$ (11). The *Pathways* prediction for the maximum rate is close to the latter value and in a good agreement with our results. Between Q_A and Q_B, our calculations for the crystallographic conformation provide the maximum rate about $(10^3 \ s^{-1})$, which is substantially smaller than observed in experiments. The dynamical amplification, however, increases the rate by almost three orders of magnitude, making it close to the experimental value of about $10^6 \ s^{-1}$ (11). At the *Pathways* level, the rate is strongly overestimated $(10^{11} \ s^{-1})$, mostly because of neglecting the destructive interference around the hydrogen bonds and around the iron ion. As expected, the *Pathways* method provides reasonable predictions for bridges with a single dominant pathway but breaks down if there is a destructive interference.

Conclusions

Effects of nuclear dynamics on electron tunneling in redox proteins have been an important question for the biological electron transfer community. While it has been understood how nuclear dynamics controls the Franck-Condon factor, little was known until now about how the dynamics affects the tunneling matrix element. Our results show that, when tunneling is dominated by a single pathway tube, dynamical effects are small and *Pathways* level calculations provide reasonable results. The situation changes when several pathway tubes are important and destructive interference exists among them. In this case dynamic amplification becomes important, i.e., tunneling is dominated by protein configurations far from the "frozen" where the destructive interference is substantially reduced. The tunneling mechanism is determined by the competition between the occupation probability of these configurations and their ability to reduce destructive interference.

We thank George Feher and Mel Okamura for valuable discussions. Work at UCSD was funded by the National Institutes of Health (Grant No. GM48043).

References

1. Bendall, D. S., ed. *Protein Electron Transfer*; BIOS Scientific Publishers: Oxford, UK, 1996.

2. Lippard. *Principles of Bioorganic Chemistry*; University Science Books: Mill Valley, CA, 1994.

3. Sigel, H.; Sigel, A. *Electron Transfer in Metalloproteins*; Marcel Dekker: New York, 1991.

4. Marcus, R. A.; Sutin, N. *Biochim. et Biophys. Acta* **1992**, *811*, 265.

5. Betts, J. N.; Beratan, D. N.; Onuchic, J. N. *J. Am. Chem. Soc.* **1992**, *114*, 4043.

6. Regan, J. J.; Onuchic, J. N. *Adv. Chem. Phys.* **1999**, *107*, pt.2, 497.

7. Gray, H. B.; Winkler, J. R. *Annual Rev. Biochem.* **1996**, *65*, 537.

8. Aquino, A.; Beroza, P.; Reagan, J.; Onuchic, J. N. *Chem. Phys. Let.* **1997**, *275*, 181.

9. Wolfgang, J.; Risser, S. M.; Priyadarshy, S.; Beratan, D. N. *J. Phys. Chem.* **1997**, *101*, 2986.

10. Daizadeh, I.; Medvedev, E. S.; Stuchebrukhov, A. A. *Proc. Natl. Acad. Sci.* **1997**, *94*, 3703.

11. Feher, G.; Allen, J. P.; Okamura, M. Y.; Rees, D. C. *Nature* **1989**, *339*, 111.

12. Stowell, M. H. B. *et al*, *Science* **1997**, *276*, 812.

13. Graige, M. S.; Feher, G.; Okamura, M. Y. *Proc. Natl. Acad. Sci.* **1998**, *95*, 11679.

14. McMahon, B. H.; Muller, J. D.; Wraight, C. A.; Nienhaus, G. U. *Biophys. J.* **1998**, *74*, 2567.

15. Hyslop, A. G.; Therien, M. J. *Inorg. Chim. Acta* **1998**, *275-276*, 427.

16. Alexov, W. G.; Gunner, M. R. *Biochemistry* **1999**, *38*, 8253.

17. Bixon, M.; Jortner, J. *J. Chem. Phys.* **1997**, *107*, 5154.

18. Kuznetsov, A. M.; Ulstrup, J. *Spectrochim. Acta Pt. A* **1998**, *54*, 1201.

19. Löwdin, P.-O. *J. Math. Phys.* **1962**, *3*, 369.

20. Priyadarshy, S.; Skourtis, S.; Risser, S. M.; Beratan, D. N. *J. Chem. Phys.* **1996**, *104*, 9473.

21. Balabin, I. A.; Onuchic, J. N. *J. Phys. Chem.* **1996**, *100*, 11573.

22. Balabin, I. A.; Onuchic, J. N. In: Proc. of the Third International Symposium on Biological Physics 1998, Santa Fe, New Mexico, USA. American Institute of Physics: 175 (1999).

23. Yates, H. *Huckel Molecular Orbital Theory*; Academic Press: New York, 1978.

24. Source: http://www.shef.ac.uk/chemistry/web-elements.

25. Naray-Szabo, G. *Applied quantum chemistry*; Dordrecht: Boston, 1987.

26. Dauber-Osguthorpe, P. et al. *Proteins: Structure, Function and Genetics* **1988**, *4*, 31.

27. Balabin, I. A.; Onuchic, J. N. *J. Phys. Chem.* **1998**, *102*, 7497.

28. Curry, B, W. *et al J. Bioenerg. Biomembr.* **1995**, *27*, 285.

29. Page, C. C.; Moser, C. C.; Chen, X.; Dutton, P. L. *Nature* **1999**, *402*, 47.

Chapter 8

Ab Initio Calculations of Long-Distance Electron Tunneling in Proteins with the Method of Tunneling Currents

Jongseob Kim, Xuehe Zheng, Yuri Georgievskii, and Alexei A. Stuchebrukhov

Department of Chemistry, University of California, Davis, CA 95616

The method of tunneling currents for the description of long-distance electron tunneling in complex molecular structures, such as proteins, is discussed. Using this method, ab initio calculations have been carried out on a model electron transfer system in which the electron donor is the blue copper center, the bridge is a polypeptide (5 glycine residues), and the acceptor is -HisRu(III)bpy$_2$Im complex. This system is a realistic model of a tunneling path in Ru-modified protein azurin studied by Gray and co-workers recently. It is shown that the tunneling matrix element as small as 10^{-4} cm^{-1} can be reliably evaluated. In the calculation, all electrons are taken into account at the Hartree-Fock level. It is shown that the tunneling process can be described using an effective one-electron picture, in which only one special pair of tunneling orbitals is considered.

1 Introduction

We have recently developed a method for the description of long-distance electron tunneling in complex molecular systems, such as proteins, which can be implemented in first principles all-electron calculations[1]-[6]. The idea of the method is to examine quantum mechanical currents induced in the system during the tunneling transition. The method proved to be useful for the description of the tunneling pathways, and provides a powerful approach for calculation of exceedingly small tunneling matrix elements in electron Donor-Bridge-Acceptor systems, where the distance between donor and acceptor is in the range of 15 to 30 Angstroms. Such systems are typical in biological electron transfer[7, 8, 9, 10], and have been the subject of active discussion in the literature in the past few years, see e.g. [11]-[17].

In this paper we briefly review the method, and describe tunneling calculations on a model Ru-modified azurin system studied by Gray and co-workers[10].

2 Tunneling currents in a tunneling transition

The method of tunneling currents has been recently reviewed[1]. Below we briefly summarize the key ideas of the approach.

We consider a system at the transition state of ET reaction, i.e. when the diabatic electronic states $|D>$ and $|A>$, which correspond to the localization of a tunneling electron on the donor and acceptor complexes respectively, are in resonance[14]. The idea of the method is to examine the spatial distribution of quantum mechanical current induced in a tunneling transition. Electron tunneling is associated with redistribution of charge in the system, therefore a tunneling transition can be characterized by the transition current.

The current density have been found to have the following form:

$$\vec{J}(\vec{r}) = -i < A|\hat{\vec{j}}(\vec{r})|D > \qquad (1)$$

where $\hat{\vec{j}}(\vec{r})$ is the current density operator,

$$\hat{\vec{j}}(\vec{r}) = \frac{\hbar}{2mi} \sum_{i=1}^{N} [\delta(\vec{r} - \vec{r}_i)\nabla_i - \nabla_i^+ \delta(\vec{r} - \vec{r}_i)], \qquad (2)$$

where the summation is over all electrons, and the hermitian conjugated operator ∇^+ is assumed to be acting on the left. Or, if the second-quantization $\hat{\psi}, \hat{\psi}^+$ operators are used, the current density has a familiar form:

$$\hat{\vec{j}}(x) = \frac{\hbar}{2mi}(\hat{\psi}_\sigma^+(x)\nabla\hat{\psi}_\sigma(x) - \hat{\psi}_\sigma^+(x)\nabla^+\hat{\psi}_\sigma(x)). \tag{3}$$

where a summation is assumed over the repeating spin index σ. Thus, to find the distribution of the tunneling currents in the system $\vec{J}(\vec{r})$ one needs to evaluate the matrix element in Eq. (1) for two diabatic states $|D>$ and $|A>$. In one electron description, the expression is reduces to a familiar form:

$$\vec{J}(\vec{r}) = \frac{\hbar}{2m}(\psi_D \nabla \psi_A - \psi_A \nabla \psi_D) \tag{4}$$

where ψ_D and ψ_A are the wave functions of diabatic donor and acceptor states. The evaluation of the current for many-electron states is more complicated and it is the subject of the later discussion.

In a slightly different formulation, instead of current density $\vec{J}(\vec{r})$, the tunneling transition can be characterized by quantum fluxes between atoms of the medium, J_{ab}, which are called interatomic tunneling currents[2]. In one-electron formulation, the interatomic currents can be expressed as follows:

$$J_{ai,bj} = \frac{1}{\hbar}(H_{ai,bj} - E_0 S_{ai,bj})(C_{ai}^D C_{bj}^A - C_{ai}^A C_{bj}^D) \tag{5}$$

where ai and bj are indices of two atomic orbitals on atoms a and b, H and S are the Hamiltonian and overlap matrices, and C^D and C^A are the coefficients of expansion of states $|D>$ and $|A>$ in the atomic basis set of the system. The total interatomic current between two atoms, J_{ab}, is a sum of $J_{ai,bj}$ over orbitals of these atoms.

The expressions for interatomic currents in many-electron formulation, unfortunately, are not as simple as Eq. (5), because of the exchange and overlap effects. (To define an atom, one needs to deal with the non-orthogonality of atomic orbitals of its neighbours, which complicates the formalism [6].)

The total current through a given atom is proportional to the probability that a tunneling electron will pass through this atom during the tunneling jump. While the matrix elements J_{ab} are proportional to probability that the tunneling electron will pass a given pair. Both the interatomic currents J_{ab} and the current density can be utilized for visualization of the tunneling process and tunneling pathways. For example, the total current through an atom can be taken as an indicator that the atom is involved in the tunneling process, see eg. Refs. [3, 4, 32], and Fig. 2 below.

The information about all tunneling paths and their interferences is contained in matrix J_{ab}, which describes the total tunneling flow in an atomic representation. The analysis of the tunneling flow gives a rigorous description of where electronic paths are localized in space. For example, if a specific atomic path exists, one can find it using the method of steepest descent, that is: begin from a donor atom, d and find an atom $b1$ to which the current $J_{b1,d}$ is maximum, then go to atom $b1$, repeat the procedure and find atom $b2$, etc., until the acceptor atom a is reached. The sequence of atoms: $d, b1, b2, ...bn, a$ is the tunneling path. Of course, this procedure will work only if a single atomic path exists. Usually, the structure of the tunneling flow is more complicated and many interfering paths exist simultaneously. A more careful analysis of J_{ab} is required in this case.

Remarkably, both $\vec{J}(\vec{r})$ and J_{ab} turns out to be related to the tunneling matrix element[2]-[6]:

$$T_{DA} = -\hbar \sum_{a \in \Omega_D, b \notin \Omega_D} J_{ab} = -\hbar \int_{\partial \Omega_D} (d\vec{s} \cdot \vec{J}). \qquad (6)$$

In the above formula Ω_D is the volume of space that comprises the donor complex, and $\partial \Omega_D$ is its surface. These relations obtained using the conservation of charge argument.

The diabatic states, by their nature, are well defined only in the regions of localization of charge, and perhaps in the tunneling barrier, but not in the far region of the other redox site. As a result, if a global property is evaluated (as opposed to local property such as current density in a given spatial point) as a matrix element over such states, such as $<A|H|D>$, in the volume integral representing the matrix there are major regions (around donor and acceptor) where one function is well defined, but the other is not. This can potentially lead to a numerical problem for extended systems, where the transfer matrix element is very small. The above formulation avoids this problem, since the expression involves the wave functions in the region of the barrier only, where *both* functions D and A are well defined.

2.1 Calculation of current density $\vec{J}(\vec{r})$

In principle, using the above formalism the current density and the transfer matrix element can be evaluated at any level of the electronic structure calculations. Here we consider the Hartree-Fock calculations.

Suppose states $|D>$ and $|A>$ are one-determinant many-electron functions, which are written in terms of (real) molecular orbitals $\varphi_{i\sigma}^D$ and $\varphi_{i\sigma}^A$, where σ is the spin index, $\sigma = \alpha, \beta$. These are the optimized canonical orbitals obtained from Hartree-Fock calculations of states D and A. Using the standard rules of matrix element evaluations[18], one can obtain an appropriate expression for Eq. (1) in terms of MO's of the system.

A key step in theory, however, is to use instead of canonical orbitals the so-called corresponding orbitals[19, 20], which are obtained in biorthogonalization [21]-[23] of $|D>$ and $|A>$ states. In ET problem such orbitals have been utilized in the past by Newton[24], Friesner[25], and Goddard[26] and their co-workers to evaluate the Hamiltonian matrix element $<A|H|D>$. It first may look just like a convenient technical step - after all the matrix element Eq. (1) can be evaluated with canonical orbitals as well, but in fact, this step provides the major insight into how one can think about tunneling in a many-electron system.

Suppose by an appropriate rotation of two sets of orbitals φ_i^D and φ_i^A they are made "pairwise orthogonal"[21]-[23]. In this case the overlap matrix of states A and D is diagonal,

$$<\varphi_{i\sigma}^A|\varphi_{j\sigma}^D> = \delta_{ij}s_i^\sigma. \tag{7}$$

and the expression for the current takes the following form:

$$\vec{J}(\vec{r}) = -\frac{\hbar}{2m}det(\mathbf{S}_{AD})\sum_{ij,\sigma}\frac{1}{s_i^\sigma}(\varphi_{i\sigma}^A(\vec{r})\nabla\varphi_{i\sigma}^D(\vec{r}) - \varphi_{i\sigma}^D(\vec{r})\nabla\varphi_{i\sigma}^A(\vec{r})) \tag{8}$$

This expression is an obvious generalization of one-electron picture. Now different pairs of corresponding (overlapping) orbitals of donor and acceptor states contribute to current density. The smaller the overlap between corresponding orbitals in donor and acceptor wave functions (i.e. the greater the change of an orbital in D and A state), the greater the contribution of a given pair to the current. All electrons in the system give direct or indirect contribution to the current, either due to proper electron/hole tunneling, or due to exchange and relaxation effects. All such contributions are in the above formula.

The pairing of the orbitals leads to a crucial physical insight since only one pair of the corresponding orbitals contributes mostly to the tunneling transition, the rest participate as an electronic Frank-Condon factor [24], as described below.

Surprisingly, the shift of the core orbitals in donor and acceptor states does not appear to be significant, i.e. their Franck-Condon factor is of the order of unity (typically 0.6 to 0.8 see next section). This is a surprising result since the *canonical* orbitals of the core electrons change significantly in donor and acceptor states [24], which is in line with a significant redistribution of the charge.

The calculation in which only one pair of corresponding orbitals is significant will be called One Tunneling Orbital (OTO) Approximation. In this approximation, the tunneling current density takes the form:

$$\vec{J}(\vec{r}) = -\frac{\hbar}{2m} <D|A>^{(0)} (\varphi_0^A(\vec{r})\nabla\varphi_0^D(\vec{r}) - \varphi_0^D(\vec{r})\nabla\varphi_0^A(\vec{r})) \qquad (9)$$

where

$$<D|A>^{(0)} = \prod_{i\neq 0} s_i. \qquad (10)$$

Except for the FC factor, the above formula coincides with that of one-electron description. In contrast with the true one-electron picture, however, all polarization (FC factor) and, and most importantly, exchange effects are present in it, not to mention the SCF manner in which MO's are found.

Before we proceed with applications, one additional remark is needed about the One Tunneling Orbital Approximation.

3 The OTO Approximation and the limitation of HF description of many-electron tunneling

The expression for currents, and therefore for the matrix element T_{DA}, Eq. (6), is given by the sum over pairs of corresponding orbitals, i=0,...N. Each of the terms has the following structure:

$$J_i = <D|A>^{(i)} <\varphi_i^D|\hat{J}|\varphi_i^A>, \qquad (11)$$

where the first factor is the product of pair-wise overlaps of all orbitals except for the i-th one, and the second factor is the matrix element taken over the i-th pair of orbitals. As is seen, the first factor is an electronic analog of the Frank-Condon overlap, which is given here by

$$<D|A>^{(i)} = \prod_{j\neq i} <\varphi_j^D|\varphi_j^A>. \qquad (12)$$

For this form of the matrix element to be correct, a specific separation of dynamic time-scales should exist in the system. Namely, as the above expression (11) suggests, the interaction associated with mixing of the i-th pair of orbitals should be much weaker (and therefore slower) than that of other orbitals. Such is the case, for example, in the non-adiabatic proton transfer, where the transfer matrix element has exactly the same form[28]:

$$T_{DA} = <\chi_i|\chi_f><i|V|f>. \qquad (13)$$

Here the first term is the overlap of vibrational functions, and the second term is the matrix element of electronic interaction V between two electronic states. The above expression is obtained only for a weak (non-adiabatic) electronic interaction[28]. The mixing of two vibrational states χ_i and χ_f in the same electronic state would occur on a much faster scale, compared to that of the weak interaction V.

Taking this analogy literary, one has to assume that in the above expression for current Eq. (11), the mixing of the i-th pair of orbitals $|\varphi_i^D>$ and $|\varphi_i^A>$ should be much slower than that of the other orbitals. In the HF SCF picture, however, all orbitals are equivalent, and the expression for current is symmetric with respect to the index of the orbitals pair i, Eq. (1). This symmetric form of the expression therefore can not satisfy the requirement of time-scale separation for *all* orbitals. This is obviously a general limitation of the SCF procedure.

In our case, there is one special pair of tunneling orbitals which indeed has the time-scale of mixing much different from the rest of the system. This can be seen from the degree of the overlap of the corresponding orbitals, as shown in Table 1. Only one pair of orbitals, $|\varphi_0^D>$ and $|\varphi_0^A>$, correspond to significant redistribution of charge in the system and therefore has a small overlap, the rest of the orbitals experience only a weak polarization shifts upon the change of the electronic state and therefore have close to unity overlaps. For the tunneling pair of orbitals the requirement of time-scale separation is satisfied, and the contribution has the expected form of Eq. (13):

$$J_0 = <D|A>^{(0)} <\varphi_0^D|\hat{J}|\varphi_0^A>. \quad (14)$$

For all other pairs of orbitals, the separation of time-scales is not satisfied. On this basis we conclude that all other terms in the current expression Eq. (8) should be simply dropped, in particular when they are small. Or, if they can not be dropped, because of their large contribution, in such cases the HF description is not applicable at all.

Below we consider specific applications, and some technical issues related to bi-orthogonalization procedure. In practical calculations, one express the MO's in terms of the atomic basis functions, and calculations are performed as described, for example in Ref. [29]. Some specific calculations will be described in the following sections.

4 Electron transfer in Ru-modified azurin. Protein Pruning.

We first discuss calculations for a model system which was motivated by experimental work of Gray and co-workers. They synthesized and studied electron transfer in Ru-modified azurin, in which His residue, introduced via site-directed mutagenesis on the surface of the protein, is coordinated to -$Ru^{3+}bpy_2Im$ complex. In Fig. 1 crystallographic structure of one of such systems is shown. Electron transfer occurs between the reduced Cu(I) center and oxidized Ru(III) complex. The distance between Ru and Cu atoms in the system shown in Fig. 1. is about 27 angstroms. The reaction is initiated by a quick laser-induced oxidation of the Ru(II) complex in the fully reduced (Cu(I)...Ru(II)) system. The electron transfer reaction is monitored in real time by observing the change in the absorption of reduced/oxidized forms of both Cu and Ru complexes. The Ru complex could be attached at different positions, therefore the distance dependence of the reaction rate could be studied in these experiments. This distance dependence is believed to be controlled primarily by the electronic coupling between Ru and Cu centers. The latter is the focus of the present study.

In the analysis of ET in proteins, the first step is to find amino acids that are involved in electron propagation in the protein medium. This can be done with a method of protein pruning which has been developed earlier in this group[33], [31]. The calculations are performed at this stage of the analysis at a semiempirical level, but has no limitation on the size of the protein. The idea is to probe the sensitivity of electronic coupling to computer modified changes in the protein. Several other groups exploited this idea with slight variations in the implementation[12]. Although the electronic coupling is calculated not very accurately, due to approximations in electronic structure treatment, the method, however, can identify the amino acids that are *not* important for the coupling. These amino acids can be deleted from the protein to simplify the model. So the essence of the method is to get rid of unimportant amino acids, i.e. to prune the protein, and identify the important ones. The details of such calculations have been described earlier in the literature[33], [31]. The result of pruning of His126 Ru-modified azurin are shown in Fig. 2.

The pruning procedure naturally leaves intact donor and acceptor complexes, and a number amino acids that make up the tunneling bridge which connects Ru and Cu ions. In this particular case, two stretches of the protein backbone provide the connection. The two stretches form "molecular wires" along which electron tunnels between donor and acceptor. The two wires were identified because for each one the connection to the redox site is strong on one end and weak on the other: the Met residue is more weakly coupled than the Cys residue to Cu ion, and the His residue of the Met wire is more strongly coupled to Ru than the Gln residue of the Cys wire to the Ru complex. The relative importance of these two paths can only be established in a more accurate calculation that can quantitatively correctly

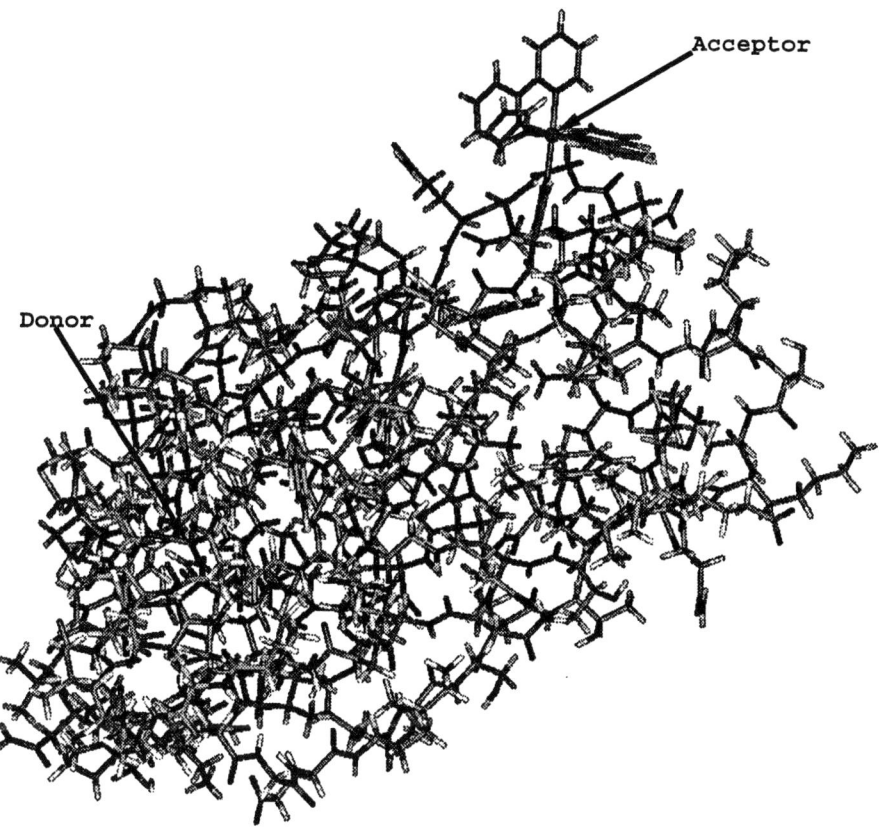

Figure 1: Ru-modified azurin system.

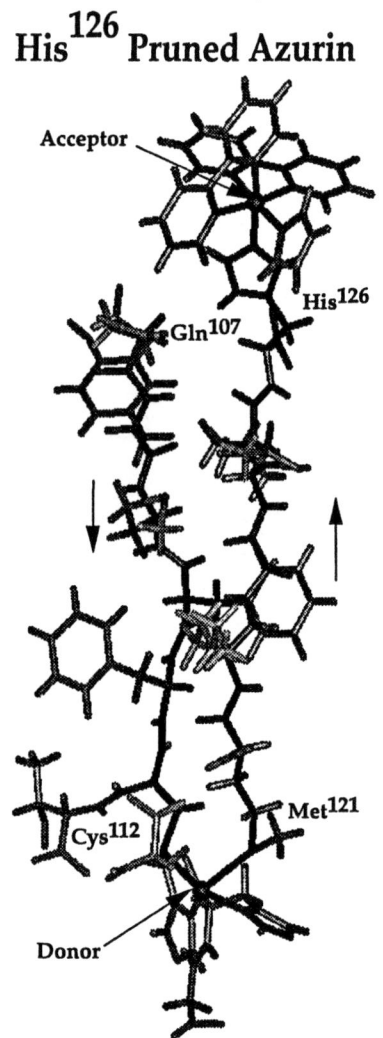

Figure 2: Pruned Ru-modified azurin system.

describe the relative strength of the weak "through space" couplings of Gln to Ru complex and Met to Cu center.

The *atoms* of the pruned molecule in Fig. 2, through which the tunneling occurs are show in various degree of dark color. Due to quality of the printing, one probably will not be able to distinguish atoms of different tint. But the method of interatomic currents, which was used for the coloring, calculates quantitatively the relative probabilities for different atoms that the tunneling electron will pass through them. The accuracy of the method is limited only by the accuracy of corresponding electronic structure calculation. As shown in Fig. 2 calculations have been done at the semiempirical Extended Hückel level. The method has been developed earlier and described elsewhere[2]-[5]. In particular in Ref. [3] a colored picture of the same system is shown, which is more informative than that shown in Fig. 2 in this issue. Other examples of the application of this method can be found, e.g., in Ref. [32].

The direction of the tunneling current, indicated by the arrows, is related to quantum mechanical phase. In our calculations all wave functions can be made real, and all the information about the quantum mechanical phase is now contained in the sign of the wave functions, which translates into the direction of the tunneling flow. The destructive interference, for example, shows up as currents in opposite directions, which can lead to cancelation, under appropriate conditions, of the currents. The total (net) flux, as we discussed in the preceding theoretical section, is related to the coupling matrix element. Thus, the mutually compensating currents imply the decrease of electronic coupling due to destructive quantum interference. Such is the case shown in Fig. 2.

The calculations shown in Fig. 2 were performed at a "low level" electronic structure description. We find, however, that the relevant part of the molecule is rather small and therefore more accurate, ab initio calculations are possible on the pruned molecule. The results of such calculations on a model system, which is derived from our pruned His126 molecule, is the subject of the rest of the paper. Our goal is to find out how exactly electrons tunnel in molecular wires shown in Fig. 2.

5 Tunneling transition in $(His)_2(Met)Cu^{1+}$-(Cys)-(Gly_5)-$(His)Ru^{3+}bpy_5Im$

This model is shown in Fig. 3. It contains two transition metal complexes, as in the pruned system, which are connected by a stretch of polypeptide chain.

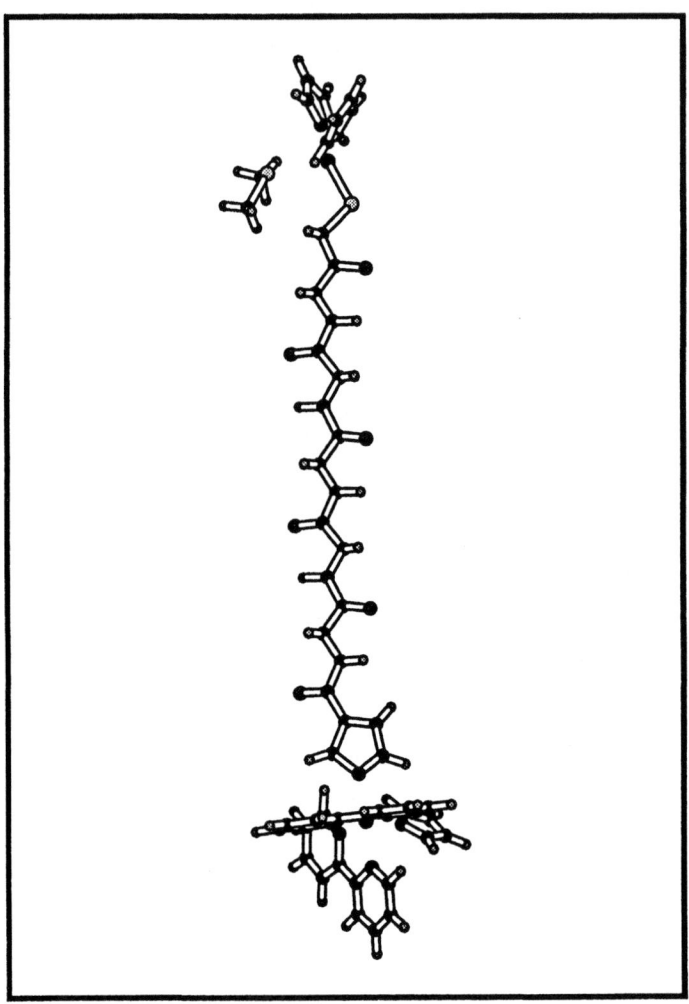

Figure 3: Model charge transfer system for which ab initio calculations described in the text, $(His)_2(Met)Cu(I/II)$-(Cys)-$(Gly)_5$-$(His)Ru(III/II)bpy_2Im$.

Electron donor and acceptor in the calculations are Cu(I) of the blue copper center in the reduced form, and the Ru^{3+} ion in -HisRu^{3+}bpy$_2$Im complex. The donor and acceptor complexes are coordinated to the opposite ends of the polypeptide (Gly$_5$) chain. The geometry of the polypeptide was chosen, for convenience of visualization, to be planar with all atoms lying in the same plane except for the hydrogen atoms of each of the α-carbons; the whole length of such a system shown in Fig. 3 is about 35 angstroms.

The two diabatic states $|D>$ and $|A>$ were calculated as described in Ref. [29, 30]. Since the two ends of the polypeptide chain are not equivalent, in the absence of an external field the donor and acceptor diabatic states, D-Bridge-A$^+$ and D$^+$-Bridge-A, have different energies, E_d and E_a, and therefore no charge transfer can occur. (E_d and E_a are total electronic energies of the system.) In order to induce the charge transfer, the two states need to be brought into resonance. In electron transfer reactions such a resonance occurs in the course of thermal fluctuations of the polar medium surrounding donor and acceptor complexes [14].

To mimic the effect of the polar environment, in our calculations an additional charge Q was placed along the axis of the molecule, and was varied to find a value at which donor and acceptor states have approximately the same energy. The same effect could be achieved by varying configuration of one or several water molecules positioned around donor and acceptor ions.

The calculations were performed by the Gaussian program with LanL2DZ basis set [34]. The structure of the tunneling flow calculated using Eqs. (8)-(9) is shown in Figs. 4a and 4b. The most prominent and typical feature in the flow is the presence of vortices, Fig. 5, similar to those reported earlier in the literature by our group in Ref. [35]. These vortices are of the same general nature as those in superfluid helium. Here, our tunneling electron can be considered as a quantum "liquid" that flows from donor to acceptor complex. The complex nature of the wave function describing dynamics, causes the appearance of quantum vertices. Further discussion of this interesting phenomenon can be found in Ref. [35].

It turns out that biorthogonalization mixes the canonical orbitals in our system in such a way that only one pair of orbitals is a major contributer to the current. We will call the orbitals of this pair φ_0^D and φ_0^A. For this pair the overlap is of the order 10^{-4}, while for others less than .1% different from 1.0. This pair of orbitals can be taken as a basis of an effective one-electron (or, one-hole) description of the tunneling process.

The structure of the tunneling flow is very important for the rate of tunneling, because it defines the magnitude of the transfer matrix element discussed in the next section.

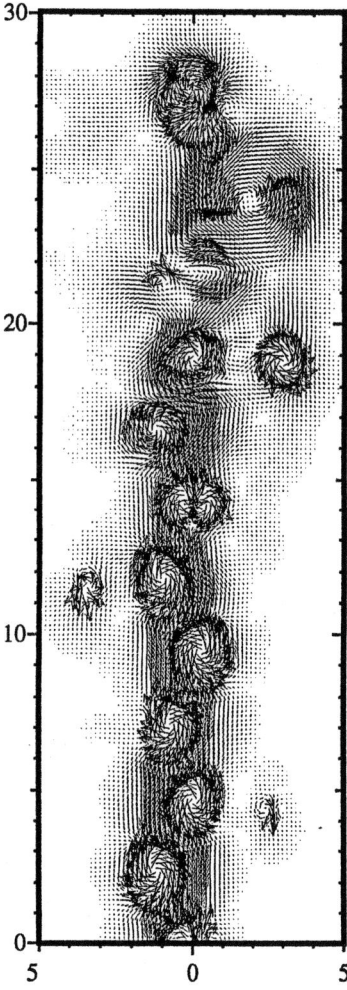

Figure 4. a. Quantum-mechanical flux in a tunneling transition in a model system shown in Fig. 3.

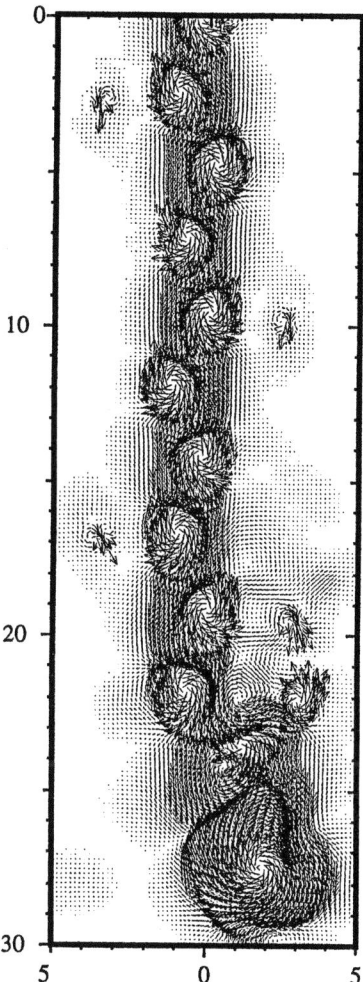

Figure 4: b. Quantum-mechanical flux in a tunneling transition in a model system shown in Fig. 3.

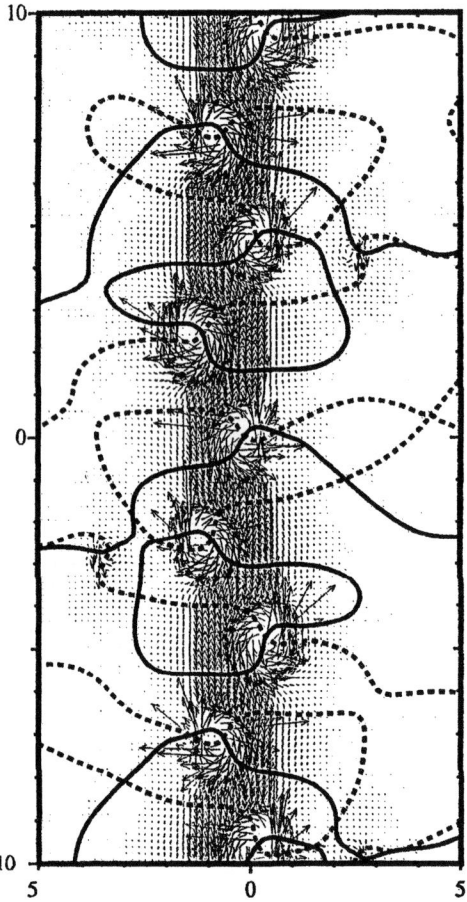

Figure 5: Middle part of the peptide system of Fig. 3 with flux and nodal lines of donor (b roken lines) and acceptor orbitals (solid lines), i.e. nodal lines of real and imaginary part s of the wave function of the tunneling electron, respectively. The crossing points of the broken and solid lin es correspond to zeros of the complex wave function of the tunneling electron and ar e the origins of the vortices.

6 Transfer Matrix Element

The transfer matrix element between donor and acceptor diabatic states could be, in principle, calculated as [15]

$$T_{DA} = H_{DA} - E_0 S_{DA}, \qquad (15)$$

where $\hat{H}$ is the Hamiltonian of the system at the transition state configuration, $\hat{H}(Q^\dagger)$, $S_{DA} = \langle D|A \rangle$ is the overlap of the two states, and $E_0 = \langle D|\hat{H}|D\rangle = \langle A|\hat{H}|A\rangle$ is their common resonance energy. The application of the above formula for large systems is difficult however. The problem is that the two terms in Eq. (15) by themselves are large numbers, of the order of 10^4 cm^{-1}, and they almost completely cancel each other since T_{DA} is typically of the order of $10^{-2} - 10^{-4}$ cm^{-1}. (In particular the resonant energy of two states is difficult to calculate accurately).

In Fig. 6 the results of calculation of the transfer matrix element by Eq. (6), which is an alternative to the above Eq. (15), are shown. The total flux was calculated through a plane perpendicular to the axis connecting donor and acceptor ions. The position of the dividing surface was varied in the range of 10 angstroms of the middle part of the bridge to check the accuracy of the calculations. It is only when the wave functions of diabatic states satisfy the Schroedinger equation in configuration space exactly, the total flux is guaranteed to be independent of the position of the dividing surface. As seen in Fig. 6, the total flux through the dividing surface fluctuates a bit, however provides an accurate estimate of the coupling matrix element. The most remarkable feature of the data in Fig. 6 is the magnitude of the flux itself, 10^{-4}cm^{-1}.

Several curves shown in Fig. 6 correspond to different transition states of electron transfer reaction. The transition state is the resonance between donor and acceptor electronic states. Such a resonance is achieved in the course of thermal fluctuations of the real system. The fluctuations of the polar medium cause the shifts of the potentials of the donor and acceptor complexes. Since fluctuations are random, the two off resonant levels can move to resonance by moving one level up, or, by moving another level down, or moving one level up and another, simultaneously, down, until their energies match. All such cases can be realized in reality, so the transition state is not uniquely defined. The question is how different these possible transition states in a real system can be? It is clear that in all cases the position of the pair of resonating states with respect to other states in the system will be different. Different will be the barrier that electron tunnels through, and therefore different will be the coupling matrix element for each individual transition state.

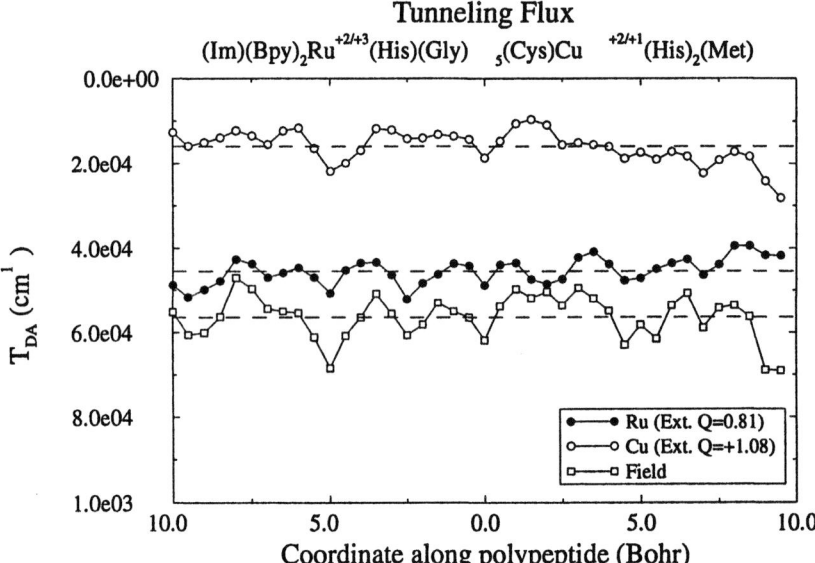

Figure 6: Total flux through a dividing surface between donor and acceptor (a plane perpendicular to the longest axis of the molecule) as a function of the position of the surface. The horizontal axis indicates the coordinate along the longest axis of the molecule, in Bohrs, as in Fig. 4, where the surface crosses the polypeptide. The vertical axis is the total flux, which by Eq. (6) is equal to transfer matrix element, measured here in wavenumbers. Different curved correspond to different transition states, described in the text.

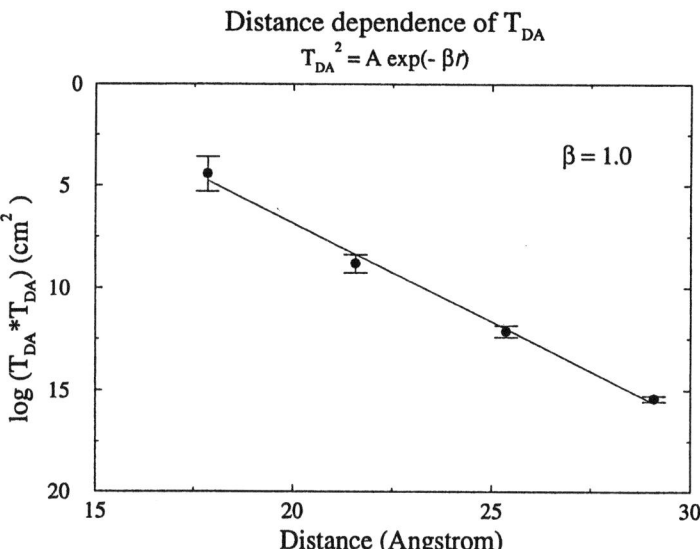

Figure 7: Distance dependence of the tunneling matrix element.

We modeled three different ways of bringing two levels in resonance, as discussed above, by varying the external charges and external electric fields. Fig. 6 shows how different the coupling matrix element in these cases might be.

For our system, the coupling surprisingly only changes maximum by factor of three. These variations should be appropriately taken into account in the rate expression, by averaging the square of the coupling matrix element, to which the rate is proportional. Additional effects due to dynamic nature of the protein medium are discussed elsewhere [36]-[38].

The distance dependence of the tunneling matrix element was examined by calculating systems with different number of Gly segments in the bridge. The dependence in this case, as expected, is exponential with β value of about 1Å^{-1}, Fig. 7. This is exactly what is observed in experiments, [10].

In our recent work, the method of tunneling currents has been implemented using semi-empirical ZINDO program[30]. The application of One Tunneling Orbital approximation made possible implementation of the interatomic currents at many-electron SCF level. In particular, the pruned system shown in Fig. 2 have been examined with this method, and the tunneling from one strand to the other via inter-strand hydrogen bonds has been investigated. The ZINDO implementation bridges the gap between accurate but demanding ab initio methods, such as discussed above, and simple but effective one-electron methods, such as the extended Hückel method.

7 How different is the Tunneling Orbital from the canonical HOMO?

One of the major findings is that a single pair of biorthogonalized orbitals forms a physical basis for describing the tunneling transfer. How different is this orbital from a the canonical HOMO? In a naive Koopmans' picture, the tunneling orbital should be HOMO. To investigate this question we have performed a series of calculations on Heme a of cytochrome c oxidase, Fig. 8. The calculations were carried using ZINDO/S model.

The standard bi-orhtogonalization procedure that is usually used in our calculations, involves two states, and two redox complexes. It turns out, however, that the orbital from which (and to which) the tunneling occurs can be found using a single complex. The calculations are performed as follows. First, the canonical orbitals of the two redox states of the system, Fe^{2+} and Fe^{3+}, are determined. A linear combination of canonical orbitals of Fe^{2+} complex is formed. There are N+1 canonical (occupied) orbitals.

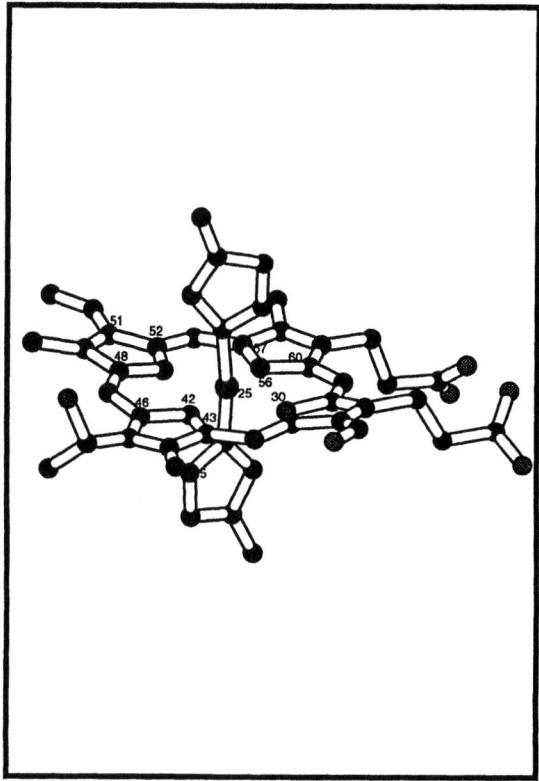

Figure 8: The model heme a system which was used to examine the nature of the tunneling orbitals.

It is then requested that the combination is such that the formed combination is orthogonal to every orbital of the Fe^{3+} complex. One can show that this procedure is equivalent to that of bi-orthogonalization carried on a system that involves both donor and acceptor complexes.

Tables 1 and 2 show the composition of HOMO of the Fe^{2+} complex, and that of the found tunneling orbital. In Table 1, we also show the LUMO of Fe^{3+} complex, as a potential "candidate" for the tunneling acceptor orbital. It is seen that both HOMO of Fe^{2+} complex, and LUMO of Fe^{3+} complex are much different from the true tunneling orbital. The degree of mixing of canonical orbitals in OTO is indicated by the cosine of the rotation, which is 0.59. A remarkable feature of the tunneling orbital, Table 2, is that it is mostly localized on the Fe atom. The HOMO, in contrast, is delocalized over several atoms of the heme.

8 Conclusion

The method of tunneling currents is a new and useful technique that allows to obtain new insights into a complex subject of tunneling in many-electron systems and provides a practical method of ab initio calculation of small tunneling matrix elements in extended systems, which are typical in biological applications. The calculations can be done on realistic models at HF level. A major finding is that the tunneling in many-electron systems typical in biological applications can be described with one pair of tunneling orbitals. These orbitals are not, however, the usual canonical HOMO orbitals. Moreover, this effective one-electron picture incorporates all the exchange and polarization effects important for tunneling. The most interesting question now is when and why the HF description of tunneling breaks down, and correlation effects become important. In particular interesting are dynamic correlations, which describe the polarization cloud around the tunneling electron/hole. Further research is needed to address these interesting issues.

9 Acknowledgments

AAS would like to acknowledge stimulating discussions of the subject of the present paper with Marshall Newton. This work was partially supported by ACS PRF, and by the National Science Foundation.

Table 1 : LUMO and HOMO of the Oxidized and Reduced Heme a:
The 5 Largest Orbital Coefficients

Heme a (Fe^{3+})	LUMO:	MO No. 148	
Atom	Atom Label	AO basis	MO Coeff.
C	52	$2p_z$	0.334
C	51	$2p_z$	0.309
N	30	$2p_z$	-0.278
C	48	$2p_z$	-0.276
C	46	$2p_z$	0.254
Heme a (Fe^{2+})	HOMO:	MO No. 148	
Atom	Atom Label	AO basis	MO Coeff.
Fe	25	$3d_{yz}$	-0.396
N	42	$2p_z$	-0.339
C	43	$2p_z$	0.305
N	56	$2p_z$	0.285
C	60	$2p_z$	-0.257

Table 2: OTO of the Oxidized and Reduced Heme a system:
The Orbital Coefficients Larger Than 0.1

Atom Label	AO basis	Atom	OTO Coeff.
5	p_x	C	0.148
25	p_z	Fe	-0.254
25	d_{xy}	Fe	-0.170
25	d_{xz}	Fe	0.120
25	d_{yz}	Fe	-0.875
42	p_x	N	-0.104
56	p_x	N	0.121
57	p_x	C	0.111

References

[1] A. A. Stuchebrukhov, Adv. Chem. Phys. **118**, 1-44, 2001.

[2] A. A. Stuchebrukhov, J. Chem. Phys. **104**, 8424 (1996).

[3] A. A. Stuchebrukhov, J. Chem. Phys. **105**, 10819 (1996).

[4] A. A. Stuchebrukhov, J. Chem. Phys. **107**, 6495 (1997).

[5] A. A. Stuchebrukhov, J. Chem. Phys. **108**, 8499 (1998).

[6] A. A. Stuchebrukhov, J. Chem. Phys. **108**, 8510 (1998).

[7] Page, C.C.; Moser, C.C.; Xiaoxi Chen; Dutton, P.L. Nature, 402: 47-52 (1999).

[8] Moser, C. C.; Keske, J. M.; Warncke, K.; Farid, R. S.; Dutton, L. P. Nature, 355: 796 (1992).

[9] Gray, H. B.; Winkler, J. R. Electron-Transfer In Ruthenium-Modified Proteins. Ann. Rev. Biochem. 65: 537-561 (1996).

[10] R. Langen, I. Chang, J. P. Germanas, J. H. Richards, J. R. Winkler, and H. B. Gray, Science **268**, 1733 (1995).

[11] Onuchic, J. N.; Beratan, D. N.; Winkler, J. R.; Gray, H. B. Electron-Tunneling Pathways In Proteins. Science, 258: 1740 (1992).

[12] Skourtis, S. S.; Beratan, D. Theories of structure-function relationships for bridge-mediated electron trasnfer reactions. Adv. Chem. Phys. 106: 377 (1999).

[13] (a) Regan, J. J.; Onuchic, J. N. Electron-transfer tubes. Adv. Chem. Phys. 107: 497 (1999). (b) Balabin, I. A.; Onuchic, J. N. Science, 290: 114 (2000).

[14] R. A. Marcus and N. Sutin, Biochim. Biophys. Acta **811**, 265 (1985).

[15] M. D. Newton, Chem. Rev. **91**, 767 (1991).

[16] Newton, M. D. Control of electron transfer kinetics. Adv. Chem. Phys. 106: 303 (1999).

[17] Newton, M. D. Modeling donor/acceptor interactions: combined roles of theory and computation. Int. J. Quant. Chem. 77: 255 (2000)

143

[18] A. Szabo and N. S. Ostlund, *Modern Quantum Chemistry* Macmillan, New York, 1982).

[19] A. Hardisson and J. Harriman, J. Chem. Phys. **46**, 3639 (1967).

[20] M. G. Cory and M. C. Zerner, J. Phys. Chem. **36**, 7287 (1999).

[21] A. T. Amos and G. G. Hall, Proc. Roy. Soc. (London) **A263**, 483 (1961).

[22] H. King, R. E. Stanton, H. Kim, R. E. Wyatt, and R. G. Parr, J. Chem. Phys. **47**, 1936 (1967).

[23] A. F. Voter and W. A. Goddard, III, Chem. Phys. **57**, 253 (1981).

[24] M. D. Newton, K. Ohta, E. Zhong, J. Phys. Chem. **95**, 2317 (1991); ibid. **92**, 3049 (1988).

[25] L. Yu Zhang, R. Murphy, and R. A. Friesner, Ab Initio quantum chemical calculation of electron transfer matrix element for large molecules. Preprint of Schrödinger, Inc. (1997).

[26] E. P. Bierwagen, T. R. Coley, and W. A. Goddard, III, in: *Parallel computing in coputational chemistry* (ACS symposium series, 592) pp 84-96.

[27] E. Heifets, I. Daizadeh, J. Guo, and A. A.Stuchebrukhov, J. Phys. Chem. **102**, 2874 (1998).

[28] Yu. Georgievskii, A. A. Stuchbrukhov, J. Chem. Phys. **113**, 10438 (2000).

[29] J. Kim and A. A. Stuchebrukhov, J. Phys. Chem. **104**, 8606 (2000).

[30] Xuehe Zheng, A. A. Stuchebrukhov, J. Phys. Chem. (2002), submitted.

[31] I. Daizadeh, J. N. Gehlen, A. A. Stuchebrukhov, J. Chem. Phys. **106**, 5658 (1997).

[32] D. M. Medvedev and A. A. Stuchebrukhov, J. Am. Chem. Soc. **122**, 6571 (2000).

[33] J. N. Gehlen, I. Daizadeh, A. A. Stuchebrukhov, and R. A. Marcus, Inorg. Chim. Acta **243**, 271 (1996).

[34] *Gaussian 94* (Revision E.2), M. J. Frisch, G. W. Trucks, H. B. Schlegel, P. M. W. Gill, B. G. Johnson, M. A. Robb, J. R. Cheeseman, T. A. Keith, G. A. Petersson, J. A. Montgomery, K. Raghavachari, M. . Al-Laham, V. G. Zakrzewski, J. V. Ortiz, J. B. Foresman, J. Cioslowski, B. B. Stefanov, A. Nanayakkara, M. Challacombe, C. Y. Peng, P. Y. Ayala, W. Chen, M. W. Wong, J. L. Andres, E. S. Replogle, R. Gomperts, R. L. Martin, D. J. Fox, J. S. Binkley, D. J. Defrees, J. Baker, J. P. Stewart, M. Head-Gordon, C. Gonzalez, and J. A. Pople, Gaussian, Inc., Pittsburgh, PA, 1995.

[35] I. Daizadeh, J. Guo, A. A. Stuchebrukhov, J. Chem. Phys. **110**, 8865 (1999).

[36] I. Daizadeh, E. S. Medvedev, and A. A. Stuchebrukhov, Proc. Natl. Acad. Sci. USA **94**, 3703 (1997).

[37] E. S. Medvedev, A. A. Stuchebrukhov, J. Chem. Phys. **107**, 6495 (1997).

[38] E. S. Medvedev, A. A. Stuchebrukhov, IUPAC **70**, 2201 (1998).

Chapter 9

Proton-Coupled Electron Transfer Reactions: A Theoretical Approach

Robert I. Cukier

Department of Chemistry, Michigan State Univerisity, East Lansing, MI 48824

Proton-coupled electron transfer (pcet) is an important mechanism for charge transfer in biology. In a pcet reaction, the electron and proton may transfer consecutively (et/pt or pt/et) or concertedly (etpt). These mechanisms are analyzed and expressions for their rates presented. Features that lead to dominance of one mechanism over another are outlined. Dissociative etpt is also discussed, as well as a new mechanism for highly exergonic proton transfer.

Introduction

The coupling between electron and proton transfer is an important pathway of charge transport in biological systems that has stimulated a great deal of theoretical (1-8) and experimental (9-14) interest. Electron transfer (et), proton transfer (pt), proton-coupled electron transfer (pcet), and dissociative pcet reactions have many features in common. They all occur by tunneling and involve large charge rearrangements with concomitant strong coupling to the surrounding medium. Deciding whether pcet is a consecutive or a concerted process can be quite difficult, from both experimental and theoretical perspectives. Important instances of pcet have product states that correspond to dissociated protons, which are able to participate in further biochemical pathways. Electron transfer can provide an initial, high-energy state for proton transfer that requires a new mechanistic explanation instead of the Marcus-Levich approach. In this article, we shall summarize our theoretical approaches to predicting mechanisms and rates of the above-noted reactions.

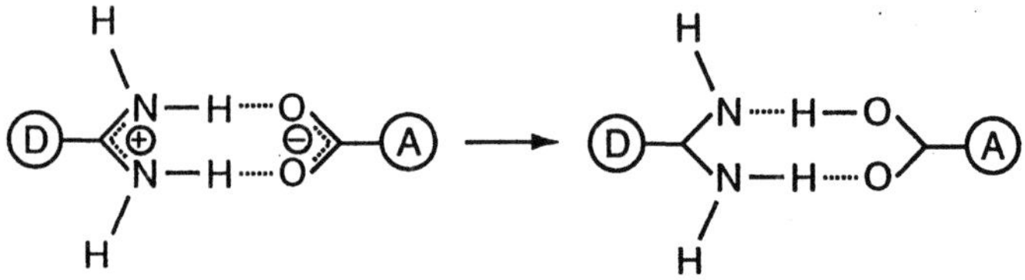

Figure 1. A molecule for a generic pcet reaction. The D and A denote electron donor and acceptor, respectively. The hydrogen bonded interface supports a proton transfer reaction. As, drawn, the hydrogen bonded interface goes from a large to a small dipole state.

A schematic of a pcet system is shown in Figure 1. The hydrogen-bonded interface provides the possibility of proton-transfer where the charge-separated interface is neutralized by (single) proton transfer. There are also electron donor (D) and acceptor (A) sites between which electron transfer may occur. The synthetic capability exists to switch the interface while keeping the orientation of the electron donor and acceptor fixed, which provides a good testing ground for theories of pcet, and such systems have been extensively studied (*12 -14*).

In the realm of biology, a step in the chain of charge transfers in the photosystem II oxygen-evolving complex (PSII/OEC) may provide an example of pcet according to the following scheme (*15*). For a *consecutive* pathway, et followed by pt, the steps are

$$P680^+ + R_1OH\text{--}NR_2 \xrightarrow{et} P680 + [R_1OH\text{--}NR_2]^{\bullet+} \xrightarrow{pt} P680 + [R_1O\text{--}HNR_2]^{\bullet+}.$$

As a concerted process, the reaction is

$$P680^+ + R_1OH\text{--}NR_2 \xrightarrow{etpt} P680 + [R_1O\text{--}HNR_2]^{\bullet+} \quad \text{Scheme I}$$

The reaction center chlorophyll P680 has been previously oxidized to $P680^+$. A tyrosine (R_1OH), labeled conventionally as Y_D, is hydrogen bonded to a nearby histidine residue (NR_2). In the electron transfer, this tyrosine is oxidized to a tryosyl radical and re-reduces P680$^+$ to P680. The proton in the tyrosyl-histidine hydrogen bond transfers from the phenol to the nitrogen of the base. The combination of et to form the tyrosyl radical and pt from the phenol to the histidine comprises the pcet system. The rate of the reaction is rather slow, around $10^3 - 10^4$ s^{-1}.

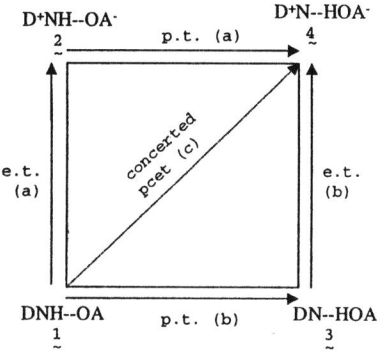

Figure 2. General scheme for pcet. The (---) lines indicate a hydrogen bond. Pathways a) and b) are consecutive schemes; respectively, et followed by pt and pt followed by et. Pathway c) is concerted etpt. If, in pathway a, et is highly endoergic, the pt step may be highly exoergic, but due to the rate limiting step principle, the overall rate will be small. The tunneling path for concerted etpt is correspondingly longer than those for individual et and pt steps, so the etpt rate constant tends to be small because of this effect.
*(Reproduced with permission from J. Phys. Chem. B, **2002**, 106. Copyright 2002American Institute of Physics.)*

Figure 2 schematizes possible charge-transfer pathways. Pathways a) and b) are consecutive and correspond, respectively, to et followed by pt, and pt followed by et, while the concerted pathway is c). In keeping with the notation in Figure 1, the electron/proton-transfer complex is denoted as DNHOA, and the charges on D, N, O, A distinguish the different possible chemical species. Which pathway will dominate is an issue of competitive rates that will be determined by electronic structure effects of the reactants and products, as well as the coupling of the electron and proton's reactant and product state charge distributions to the surrounding "solvent" that, in biological contexts, typically consists of proteins, bound and free water, and various ions and other, polar species. If the solvation energetics dictates, for example, that the et reaction in pathway a) is strongly endergonic, then it is unlikely that this will be a competitive pathway. Even if the following pt reaction is fast, consecutive reaction schemes are limited by their slow step. On the other hand, the concerted transfer of an electron and proton may be more difficult than individual transfers, as these reactions occur by the intrinsically low-probability event of quantum mechanical tunneling, and transferring two quantum particles simultaneously can be difficult.

In addition to pcet where the proton's final state is bound, it is also of interest to consider dissociative pcet, where the proton's fate corresponds to a dissociative state. It should be stressed that the consecutive reaction pathways, while corresponding to independent events from the perspective of the tunneling of the species, are actually coupled via the charge rearrangements that occur. Indeed, after electron transfer occurs in pathway a) of Figure 2, the potential subsequent proton-transfer initial state may well correspond to a poised system, comparable to a photoacid, where the proton transfer is quite exergonic. In view of this general feature, we will also discuss pure proton transfer for strongly exergonic reactions.

Proton-Coupled Electron Transfer

If the consecutive reaction pathways a) and b), indicated in Figure 2, dominate the kinetics, then the overall rate k follows the form of a rate-limiting law $k^{-1} = k_{et}^{-1} + k_{pt}^{-1}$, where k_{et} and k_{pt} are, respectively, electron- and proton-transfer rate constants. In the Marcus-Levich (2,16) non-adiabatic formulation of charge transfer, these rate constants are given as

$$k = V^2/\hbar \sqrt{\pi/\lambda_s k_B T} \, e^{-(\lambda_s + \Delta G^0)^2 / 4\lambda_s k_B T}. \tag{1}$$

The parameters in Eq. (1) that come from the coupling of the reactant and product charge distributions to the surrounding medium are the solvent reorganization (λ_s) and reaction free (ΔG^0) energies. When the reactant and product states are weakly coupled (nonadiabatic reactions), V is the electronic (protonic) matrix element V_{el} (V_{pr}). Note that the parameters λ_s and ΔG^0 depend on the initial and final charge states of the reactions; thus they can be quite distinct for the electron- and proton-transfer reactions. Furthermore, their dependencies on the initial and final charges are different, leading to distinct, differing trends in λ_s and ΔG^0 with changes in charge states.

Turning to pathway c), the concerted-reaction mechanism, we have formulated two approaches to predicting the rate constant, a double-adiabatic and a two-dimensional approach (4,5). In the double-adiabatic theory, the electron is considered to be coupled to two nuclear modes, a solvent (orientational polarization) mode that is treated classically in view of its low frequency and a bond motion, treated quantum mechanically in view of its high frequency, for the proton that is in the hydrogen-bonded interface. If, concerted with the electron transfer from D to A, the proton mode is sufficiently displaced so as to correspond to switching the hydrogen bond, as drawn in Figure 1, then, in one quantum mechanical event, both the electron and proton have transferred. A more revealing view of the process can be formulated in a two-dimensional tunneling space, whereby electron and proton are both treated as quantum objects that must both tunnel through a two-dimensional profile. This treats electron and proton on an equal footing. However, the mass disparity between the two suggests a restricted tunnel path in the two-dimensional space where the proton displaces adiabatically along its coordinate to a certain position that permits the electron to tunnel (one-dimensionally) along its coordinate. Then the proton, with the electron in its final state, relaxes to its final state. This zig-zag tunnel path is illustrated in Figure 3 where i and f denote the initial and final electron states and a and b the initial and final proton states. Either approach eventually leads to the following rate expression for concerted pcet:

$$k_{etpt} = \frac{V_{el}^2}{\hbar}\sqrt{\frac{\pi}{\lambda_s k_B T}}\sum_{n'}\rho_{in'}\sum_{n}\left|\left\langle \chi_{fn} | \chi_{in'}\right\rangle\right|^2 e^{-\left(\lambda_s+\Delta G^0+\varepsilon_{fn}-\varepsilon_{in'}\right)^2/4\lambda_s k_B T}. \quad (2)$$

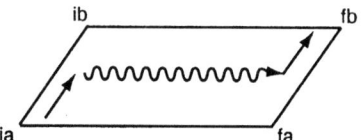

Figure 3. Zig-zag tunneling path in a two-dimensional electron-proton tunneling space. The wavy line denotes the electron's tunneling when the proton rearranges to the proper configuration to symmetrize the electron potential energy surface. The straight arrows denote the protons motion for the initial i and final f electron states.
(Reproduced with permission from J. Phys. Chem. B, 2002, 106. Copyright 2002American Institute of Physics.)

The new ingredients are Franck-Condon factors of the initial (final) state proton vibronic wavefunctions $\chi_{in'}$ (χ_{fn}). The initial states are summed over the equilibrium initial state proton distribution, $\rho_{in'}$. The "effective" activation energy appearing in the exponent of Eq. (2) involves the energetic difference of the proton eigenstates, $\varepsilon_{fn}-\varepsilon_{in'}$, and arises from the requirement of overall energy conservation between the initial and final electron-proton states. The tunneling of the proton is manifested in the Franck-Condon (FC) factors in Eq. (2). Thus, concerted pcet is limited by a "FC drag" that is a reflection of the two-dimensional tunneling requirement. The rate constants k_{et} and k_{pt} require evaluation of λ_s and ΔG^0 that depend on the specifics of the charge rearrangement and the medium. These quantities are calculable by methods based on dielectric continuum modeling, solution of the Poisson-Boltzmann equation, or molecular dynamics (MD) simulation. The electronic (protonic) matrix element V_{el} (V_{pt}) is an intrinsically quantum mechanical parameter that has to be evaluated by quantum chemical methods. Often it is simply taken as a parameter of the theory. For k_{etpt}, in addition to the indicated quantities, a model for the proton potential surfaces, when the electron is in its initial and final state, is required. Typically, these are double-well potentials that may support proton localized states on either side of the flanking heavy atoms. Then, FC factors can be obtained by evaluating $\chi_{in'}$ (χ_{fn}) and energies $\varepsilon_{in'}$ (ε_{fn}) for the proton when the electron is in its initial and final states.

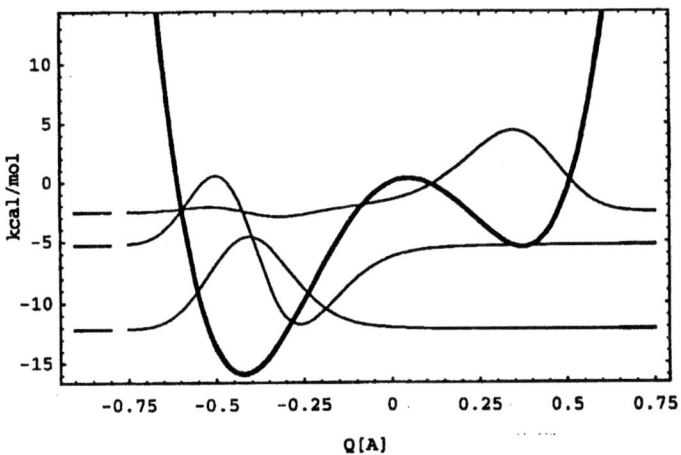

Figure 4. An initial electron state that supports both et and etpt reaction channels. The energies of the proton-localized levels are shown along with their corresponding wave functions. The ground and first excited-state wavefunctions are mostly localized in the initial state, while the second excited state is mostly localized in the final state. The pattern for the final electron state is obtained by inversion through the origin. In the initial electron state, the lowest proton (left-localized) proton state has good Franck-Condon overlap with the third proton state of the final electron state surface, and can therefore permit a high et rate constant. If the hydrogen bonding strength is increased, the potential surface will no longer have a well on the right side that can support bound proton states. In this case, only etpt can occur.

Figure 4 displays typical proton surfaces along with some of the wavefunctions and energies. It is important to note that typical proton surfaces in, e. g., amine-carboxylate hydrogen-bonded interfaces do not support many bound proton states. It also should be noted that these surfaces are not pure gas-phase surfaces; rather, they include the effects of electronic solvation from the solvent electronic degrees of freedom. Thus, charge-separated states that would not be favored in the gas phase can be greatly stabilized by such solvation. We have evaluated the rate constants, appropriate to the three paths shown in Figure 2, for a variety of potential proton-coupled electron transfers (5). While general rules are hard to provide, there is a definite tendency for etpt to be favored over et/pt or pt/et in strongly hydrogen bonded systems, and when the solvation effects are not too strong. Such situations can lead to proton states that are only available on one side of the hydrogen-bonded interface, the side being favored switching as the electron transfers. With reference to Figure 4, if the hydrogen bonding is somewhat greater than displayed there, then the potential surface will not have proton localized states on the right (left) for the initial (final) electron state. In this limiting situation, only the etpt channel is possible. On the other hand, weaker hydrogen bonding and/or strong solvation, where the proton undergoes a large change in dipole upon transfer, tends to favor the et mechanism. Assuming that the corresponding pt rate constant is not rate limiting, then this circumstance would favor et/pt versus etpt. Let us stress that it is an issue of parallel reaction channels—consecutive and concerted channels can exist simultaneously.

Dissociative Proton-Coupled Electron Transfer

We now consider reactions where the final state is dissociative, and denote the concerted process as detpt. Schematically, the reaction is

$$A + R_1OH\text{--}NR_2 \xrightarrow{detpt} A^- + [R_1O]^\bullet + [HNR_2]^+. \quad \text{Scheme II}$$

The concerted process involves the simultaneous transfer of an electron and proton, with the proton final state characterized by a repulsive potential energy surface. Experimental evidence for detpt comes from a variety of sources, though it is not easy to prove that a given process is consecutive. As noted above, a step in the chain of charge transfers in the photosystem II oxygen-evolving complex (PSII/OEC) has been postulated to occur by pcet, but there is also (17,18) a formally similar pathway along a symmetry related branch. This involves the same reactants as in Scheme I, with now the tyrosine designated as Y_Z, but there is no evidence for a hydrogen-bonded proton at a well-defined distance after proton transfer. Thus, the postulation of a dissociative step is introduced. The rate constant for this pathway is around 10^7 s^{-1}.

An extensive study, by Linschitz and coworkers (19), of the quenching of triplet C_{60} by phenols in the presence of substituted pyridines does provide unambiguous evidence for concerted detpt. Phenoxy radicals and a protonated base are the observed reaction products, in addition to $C_{60}^{\bullet-}$, demonstrating the concerted nature of the reaction. Analysis (20) of intermediates and products of radiation-induced electron-transfer reactions of substituted phenols in low-temperature glasses indicate that detpt occurs by both mechanisms.

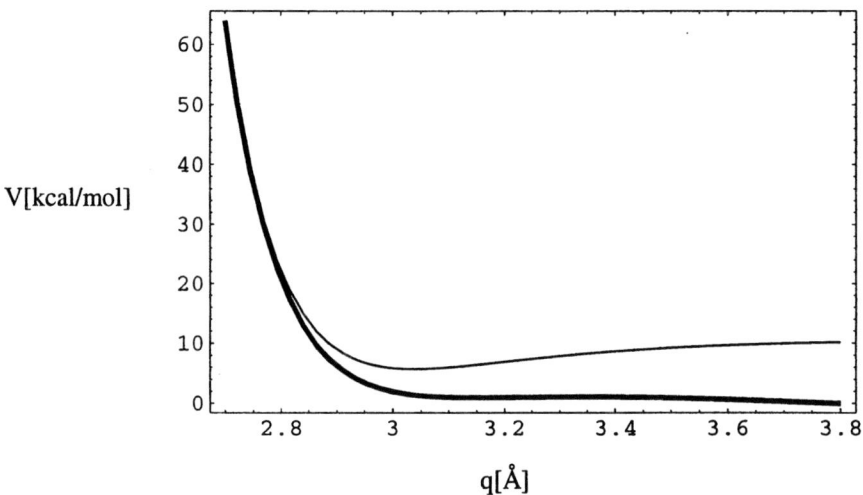

Figure 5. A gas phase surface $V_g(q)$ (solid line) and electronically solvated pels surface $V_f^{pels}(q) = V_g(q) + V_f^{els}(q)$ (bold line) as a function of the oxygen-hydrogen distance, q. The electron has transferred, and the coordinate describes the breaking of the O--H hydrogen bond. The proton electronically solvated (pels) surface is repulsive, leading to dissociative concerted electron proton transfer.
(Reproduced with permission from reference 21. Copyright 1999 American Institute of Physics.)

The double adiabatic approach provides a convenient starting point for a detpt theory (21). The principle modification is the treatment of the FC factors for the overlap of the proton initial and final eigenstates, when the final proton state is characterized by a repulsive surface. The sum over final proton states becomes an integration over a continuum of states, and bound-unbound FC factors need to be evaluated. An approach can be formulated with methods that have been used to discuss bond-breaking electron-transfer reactions (22). If the motion along the repulsive surface for the dissociation can be treated classically, then simplified expressions can be generated for the rate constant. Indeed, we have shown that the rate constant can be written in the form

$$k_{detpt} = \frac{V_{el}^2}{\hbar\sqrt{\lambda_s k_B T}} \sum_{n'=0} \rho_{1n'} \int dq \left|\chi_{1n'}(q)\right|^2 e^{-\left(\lambda_s + \Delta G^0 + V_2(q) - \varepsilon_{1n'}\right)^2 / 4\lambda_s k_B T} \qquad (3)$$

where $V_2(q)$ is the pes for the proton coordinate q. This expression has the form of a weighted sum over initial proton states of bound (initial proton states)-unbound (final proton states) FC factors with effective activation energies that reflect overall energy conservation with a continuum of final proton states.

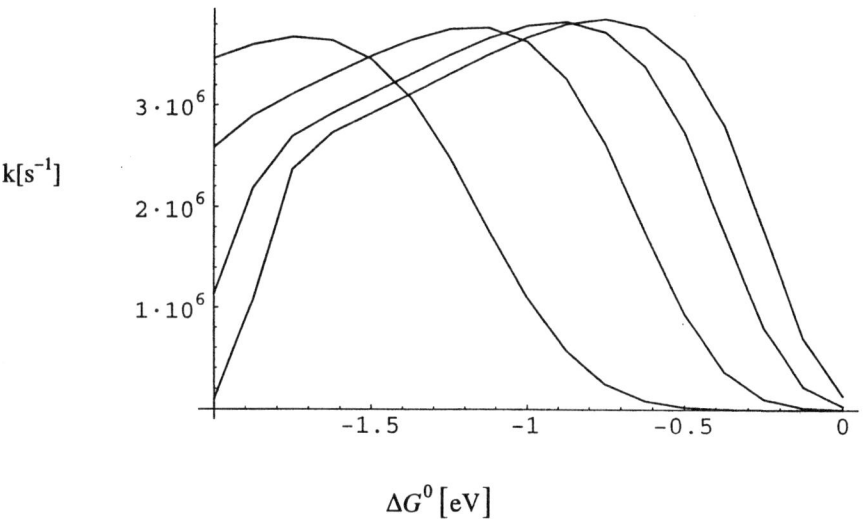

Figure 6. The detpt rate constant k as a function of ΔG^0 for λ_s=0.125, 0.25, 0.5 and 1.0 eV. As λ_s increases, the curves shift their respective maxima to more negative ΔG^0 values.
(Reproduced with permission from reference 21. Copyright 1999 American Institute of Physics.)

Dissociative proton surfaces may result from electronic structure effects after electron transfer. Another possibility arises from the effect of electronic solvation on the gas-phase surface. For example, Figure 5 shows a typical Lippincott-Schroeder gas-phase surface for a phenol amine cation radical $[R_1O\text{--}HNR_2]^{\bullet+}$ oxygen-proton hydrogen-bond stretch, along with the corresponding electronically solvated surface. The influence of electronic solvation turns the bound into a repulsive surface. It is worth noting how this comes about. Dielectric continuum arguments show that the Born solvation energy is a sum of electronic and orientational polarization contributions with roughly equal numerical contribution from each. Since Born solvation energies are large on the scale of hydrogen-bond energies, electronic solvation effects on these protonic surfaces can be significant. We will refer to these surfaces as *pels* (proton electronically solvated) surfaces, and discuss their construction with the use of MD methods in the following Section.

Using the simple exponential repulsive form $V_2^{pels}(q) = D_r e^{-\kappa q}$, suggested by the pels plot in Figure 5, along with the proton wavefunctions and energies in the (bound) initial state well, then permits evaluation of Eq. (3). In Figure 6, we display typical results for k_{detpt} in the form of "Marcus" plots. The striking feature of these results is that the magnitude of the maximum rates are all quite close to each other though, naturally, the maxima occur at differing values of ΔG^0. The similar rate maxima occur because the continuum of final proton states provides many opportunities for the quantum transition. This is in contrast to etpt, where both initial and final electron/proton states are bound, and the rate constant values depend quite sensitively on the number and energies of the final bound states.

Another interesting feature of detpt is the modest isotope effect that occurs for typical parameter values. Replacement of the proton by a deuteron is expected to reduce the rate constant, since the deuteron tunnels less than a proton, but quantitative evaluation of the proton to deuteron rate constant shows no more than about a factor of two rate enhancement of proton relative to deuteron. Such modest isotope effects are common among reactions that are thought to involve proton-coupled electron transfer.

Strongly Exergonic Proton Transfer Following Electron Transfer

As noted in the Introduction, if electron transfer is strongly endergonic, the following proton transfer may be strongly exergonic. For example, oxidation of a phenol can increase the acidity of the resulting phenoxyl radical cation by around 10 pK units, corresponding to a ΔG^0 of ~ –14 kcal/mol. That is similar to what occurs in photoacids, where the excited state acidity is much greater than that of the ground state. In the view of proton transfer that is based on the Marcus-Levich electron transfer picture (*2,16*), the transition state for the transfer is found parametric on the solvent's configuration; it corresponds to solvent fluctuations that provide equality of energy for the reactant and product state proton configurations. The symmetrized proton surface permits the proton to transfer by resonance, and this point of view for a weak overlap case leads to the rate constant expression of Eq. (1).

For a strongly asymmetric proton surface, the solvent's symmetrizing ability may not be great enough to occur with sufficient probability to give a measurable rate. The symmetrizing parameter is given by $\Delta = \sqrt{2\lambda_s k_B T}$; it measures the scale of fluctuations to be expected in the proton surface by its coupling to the solvent. (Using $\lambda_s = 1600$ (6400) cm^{-1} = 0.2 (0.8) eV gives $\Delta = 800$ (1600) cm^{-1} ~ 2.3 (4.6) kcal/mol.) These are not large numbers on the scale of asymmetries appropriate to highly exergonic reactions. Nevertheless, the proton may transfer at a robust rate based on the following mechanism that drives the proton transfer but does *not* involve solvent symmetrization of the proton pes (23). Since the proton is fast, relative to the solvent degrees of freedom, the proton surface should be evaluated parametric on the solvent's configuration, **R**. We then introduce a proton-transfer rate-constant parametric on the solvent configuration, $k(\mathbf{R})$, and construct the rate constant as an average over an appropriate distribution of solvent configurations. The $k(\mathbf{R})$ values can be evaluated by, e. g., WKB methods.

The appropriate proton potential surface parametric on **R** was constructed as follows. We used a gas-phase surface that was fit to an *ab initio* study of a phenol-ammonia H-bonded complex in its first excited singlet state (24). It is electronically solvated by using a model of dichloromethane with inducible dipoles centered on sites in the dichloromethane molecules. These degrees of freedom fluctuate on the electronic time scale, so they are much faster than the proton's time scale, and the resulting pels surface can be viewed as a potential of mean force. This surface is then coupled to the nuclear degrees of freedom of the solvent by conventional MD. The nuclear degrees of freedom correspond to the orientational polarization degrees of freedom in a dielectric continuum picture. They are slow, compared with the proton's motion, since characteristic solvent frequencies for orientational motion is in the 1-100 cm^{-1} range while the proton's frequency in the hydrogen bond is 2000-2500 cm^{-1}.

In Figure 7, we show a set of solvated proton surfaces taken from this MD simulation. The two horizontal lines are the reactant and product energy minima of the pels surface. There are a sufficient number of surfaces displayed to assess the bandwidth (Δ ~5 kcal/mol) that the nuclear degrees of freedom of the solvent impose on the pels surface. Since the asymmetry of the pels surface is about 15 kcal/mol, it would be a very rare fluctuation, indeed, that would symmetrize the surface to permit a Marcus-Levich reaction channel. The data of Figure 7 were generated with configurations obtained from equilibrium fluctuations of the solvent with the solute corresponding to the proton in the reactant state after electron transfer. This is appropriate to a homogeneous kinetic regime where the rate constants are smaller than the solvent relaxation time(s). This is not always the case, as we now discuss.

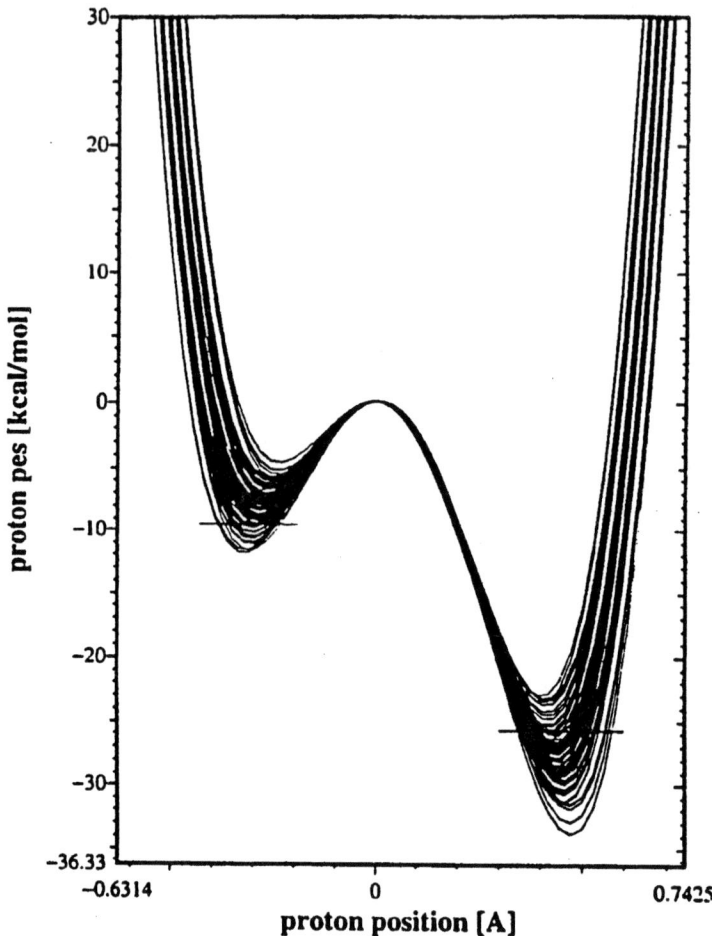

Figure 7. A set of solvent-configuration-dependent proton surfaces. There are a sufficient number of surfaces represented to represent the bandwidth that the solvent produces on the proton electronically solvated (pels) surface. The horizontal lines mark the reactant and product state minima of the pels surface. The configurations are generated with the solvent equilibrated to the proton in the electron reactant state, as appropriate to proton transfer being slow relative to solvent relaxation.

(Reproduced with permission from reference 23. *Copyright 1999American Institute of Physics.)*

For fast proton transfer relative to the solvent's response to the charge change that accompanies the Franck-Condon excitation corresponding to the electron's transfer, the solvent cannot equilibrate to the new solute charge distribution on the proton-transfer time scale. Then, the solvent configuration average should be carried out using configurations appropriate to the electron being in its initial state. Conversely, if proton transfer is slow compared to the solvent response, configurations sampled from equilibration to the charge distribution appropriate to the final electron state should be used (as used to generate Figure 7). In general, when there is a configuration-dependent rate process coupled to a relaxation (solvation) process, the macroscopic kinetics are inhomogeneous. In this case, a survival probability scheme appropriate to inhomogeneous kinetics needs to be used. Inhomogeneous kinetics will occur only for very fast kinetics where proton transfer is so fast that it competes with solvent relaxation. For proton transfers that are slower than, e. g., 5 ps^{-1} in polar solvents that have fast relaxation in response to charge changes, the kinetics should be homogeneous. This is, indeed, what we have found.

Rate constants, based on potential profiles such as those displayed in Figure 7, were evaluated with use of the WKB expression. The ingredients for the evaluation of $k(\mathbf{R})$, the barrier height above the reactant well eigenvalue and the well frequency, were obtained by solution of the Schrödinger equation for these potential surfaces. We also developed a survival probability MD scheme to account for the interference between solvation dynamics and proton transfer, and then compared the two predictions. The survival probability decays reasonably exponentially. The rate constant obtained from the survival probability scheme is about 50% smaller than the rate constant $<k(\mathbf{R})>$ obtained by direct configuration averaging. The value of $<k(\mathbf{R})> \sim 0.25$ ps^{-1} and shows that this proton-transfer channel does lead to very fast rates. Such a fast rate could not be obtained from the Marcus-Levich mechanism for this degree of surface asymmetry.

Concluding Remarks

In this symposium contribution we focused on pcet where the electron and proton dynamics are intimately coupled. Concerted pcet—etpt and detpt—were formulated as two-dimensional tunneling reactions, where the electron and proton are treated on an equal footing, driven by coupling to a solvent coordinate. A new mechanism relevant to highly exergonic proton transfer reactions was also discussed. Practical expressions for predicting rates were presented. Their evaluation should be done in the future with either Poisson-Boltzmann or molecular dynamics methods, to obtain the solvation energetics. Once MD is used to obtain solvation energetics, it can also be linked to the quantum mechanical methods used to obtain the proton energy levels and wavefunctions required for the Franck-Condon factors. This combination of methods should permit a more realistic comparison of the theoretical predictions with experiments.

Acknowledgement: The financial support of the National Institutes of Health (GM 47274) is gratefully acknowledged.

References

1) Krishtalik, L. I. *Adv. Electrochemistry Electrochem. Eng.* **1970**, *7*, 283.
2) Levich, V. G. *Physical Chemistry - An Advanced Treatise*; Henderson, H. and Yost, W., Ed.; Academic: New York, 1970; Vol. 9B, pp 985.
3) Cukier, R. I. *J. Phys. Chem.* **1994**, *98*, 2377.
4) Cukier, R. I. *J. Phys. Chem.* **1995**, *99*, 16101.
5) Cukier, R. I. *J. Phys. Chem.* **1996**, *100*, 15428.
6) Cukier, R.; Nocera, D. *Ann. Rev. Phys. Chem.* **1998**, *49*, 337.
7) Benderskii, V. A.; Grebenshchikov, S. Y. *J. Electroanal. Chem.* **1994**, *375*, 29.
8) Fang, J.-Y.; Hammes-Schiffer, S. *J. Chem. Phys.* **1997**, *107*, 8933.
9) Okamura, M. Y.; Feher, G. *Annu. Rev. Biochemistry* **1992**, *61*, 861.
10) Ferguson-Miller, S.; Babcock, G. T. *Chem. Rev.* **1996**, *96*, 2889.
11) Binstead, R. A.; McGuire, M. E.; Dovletoglou, A.; Seok, W. K.; Roecker, L. E.; Meyer, T. J. *J. Am. Chem. Soc.* **1992**, *114*, 173.
12) Turro, C.; Chang, C. K.; Leroi, G. E.; Cukier, R. I.; Nocera, D. G. *J. Am. Chem. Soc.* **1992**, *114*, 4013.
13) Roberts, J. A.; Kirby, J. P.; Wall, S. T.; Nocera, D. G. *Inorg. Chim. Acta* **1997**, *263*, 395.
14) Kirby, J. P.; Roberts, J. A.; Nocera, D. G. *J. Am. Chem. Soc.* **1997**, *119*, 9230.
15) Tommos, C.; Babcock, G. T. *Acc. Chem. Res.* **1998**, *31*, 18.
16) Marcus, R. A.; Sutin, N. *Biochim. Biophys. Acta* **1985**, *811*, 265.
17) Hoganson, C. W.; Lydakis-Simantiris, N.; Tang, X.-S.; Tommos, C.; Warncke, K.; Babcock, G. T.; Diner, B. A.; McCracken, J.; Styring, S. *Photosyn. Res.* **1995**, *46*, 177.
18) Hoganson, C. W.; Babcock, G. T. *Science* **1997**, *277*, 1952.
19) Biczok, L.; Gupta, N.; Linschitz, H. *J. Am. Chem. Soc.* **1997**, *119*, 12601.
20) Brede, O.; Orthner, H.; Zubarev, V.; Hermann, R. *J. Phys. Chem.* **1996**, *100*, 7097.
21) Cukier, R. I. *J. Phys. Chem. A* **1999**, *103*, 15989.
22) German, E. D.; Kuznetsov, A. M. *J. Phys. Chem.* **1994**, *98*, 6120.
23) Cukier, R. I.; Zhu, J. *J. Chem. Phys.* **1999**, *110*, 9587.
24) Yi, M.; Scheiner, S. *Chem. Phys. Letts.* **1996**, *262*, 567.

Chapter 10

Proton Relay in Membrane Proteins

Régis Pomès

Structural Biology and Biochemistry, Hospital for Sick Children,
555 University Avenue, Toronto, Ontario M5G 1X8, Canada

The molecular mechanism mediating long-range proton transport in hydrogen-bonded networks embedded in membrane proteins is examined. General considerations on proton mobility in proton pumps are outlined. Recent molecular dynamics studies of proton relay along chains of water molecules in the lumen of the gramicidin channel and in a non-polar analogue are reviewed.

1 Plumbing and Pumping

Lipid bilayers contain nonpolar cores made up of aliphatic chains. For that reason, they constitute microscopic capacitors that are largely impermeable to ions. Nature has harnessed this property by using the build-up of electrochemical proton gradients, $\Delta\mu_{H+}$, across biological membranes, to perform energy transduction [Mitchell, 1961]. To transport H^+ across the membrane, special molecular assemblies are required. The unproductive equilibration of H^+ across biological membranes can be accomplished by proton-conducting channels such as gramicidin [Tian and Cross, 1999; Wallace, 1998]. By contrast, in energy-transducing membranes, proton transport is coupled to chemical reactions [Saraste, 1999]. Examples of energy-transducing protein assemblies include the proton pumps bacteriorhodopsin (BR) [Lanyi, 1999] and cytochrome *c* oxidase (COX) [Wikström, 1977]. BR is a light-driven proton pump found in the purple membrane of *Halobacterium salinarum*. COX, the terminal link of the electron-transport chain in mitochondrial and bacterial respiration, is a redox-coupled proton pump. In the catalytic cycle of this enzyme, the energy released by the reduction of molecular oxygen to water is used to pump four protons against an electrochemical H^+ gradient. Subsequently, the energy gained from the downfield translocation of H^+ is harnessed by another complex transmembrane assembly, ATP synthase, to phosphorylate adenosine diphosphate (ADP).

The translocation of protons over large distances can be achieved by a Grotthuss mechanism [De Grotthuss, 1806; Agmon, 1995] involving proton-hopping along hydrogen-bonded networks embedded in membrane-spanning proteins. Such networks involve titratable amino acid residues of membrane-spanning protein assemblies as well as internal water molecules. Because of the extended and complex nature of hydrogen-bonded pathways, the molecular properties and detailed mechanisms governing H^+ translocation in proteins have remained elusive. In particular, a high level of detail is required to understand how proton pumping arises, and how it is coupled to driving forces resulting from light activation or redox reactions. In recent years, the elucidation of a number of three-dimensional structures of transmembane protein assemblies involved in proton translocation has opened the way to detailed mechanistic studies through the analysis of such hydrogen-bonded networks. The purpose of this brief overview is to provide qualitative elements for a discussion of proton relay at play in proton equilibration, proton leakage, proton uptake, and proton release mechanisms in energy-transducing membrane proteins, in the light of recent advances in the theoretical study of the Grotthuss mechanism in biological hydrogen-bonded networks. We first consider the interplay between the driving force for net H^+ transport and the transient events involved in proton pumping.

1.1 Plumbing: a role for hydrogen-bonded networks

Hydrogen-bonded networks possess a special property: they can mediate the long-range translocation of H^+ via chemical exchange of hydrogen nuclei. This process was first imagined in 1806 by De Grotthuss in relation to electrolysis of water in galvanic cells [De Grotthuss, 1806] and was formulated more recently in the context of biological systems [Onsager, 1967; Nagle and Morowitz, 1978; Knapp et al., 1980; Brunger et al., 1983]. The elementary exchange step of the Grotthuss mechanism consists of proton transfer between adjacent hydrogen-bonded groups in a hydrogen-bonded network. As depicted schematically in

Fig. 1, successive transfer steps along a suitably-oriented chain, or "proton wire," results in the net transport of one proton from end to end, without the need for a proton-carrying molecule to diffuse throughout the system. This hopping is known as the transport of an *ionic defect*. Proton hopping leaves the chain in the opposite orientation, so that in order for a second proton to be translocated in the same direction, the inversion of the chain must first take place. The reorientation of each H-bearing group in the chain creates a defect in the continuity of the hydrogen-bonded chain (HBC), so that the overall reorientation process is described as the translocation of a *bonding defect*. Both "proton-hop" and subsequent "turn" steps are thus required in the directional (vectorial) transport of protons.

1.2 How does a proton pump work?

To understand the energetics of energy-transducing proteins, it is essential to consider the relative proton affinities of the chemical groups involved in the sequence of individual proton transfer steps through the protein interior [Warshel, 1979]. A simple scheme for the energetics of proton pumping pathways is proposed in Fig.2. The diagram considers four states of a possible sequence resulting in the net transport of one proton from the side of the membrane with low proton activity (I) to the side of high activity (O) by a group located in the transmembrane region (B). Two proton-relaying groups A and C located on either side of group B are also shown for illustrative purposes. Groups A and C are among the groups that can relay H^+ with B respectively in the uptake (I side) and release (O) pathways for pumped protons.

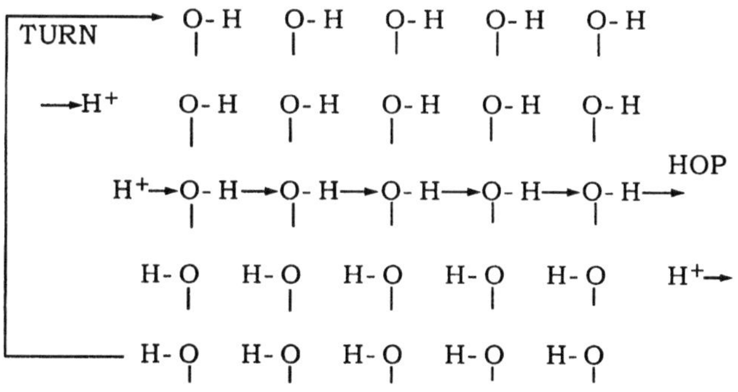

Figure 1. Schematic representation of hop and turn steps in the Grotthuss mechanism along a hydrogen-bonded chain (see text).

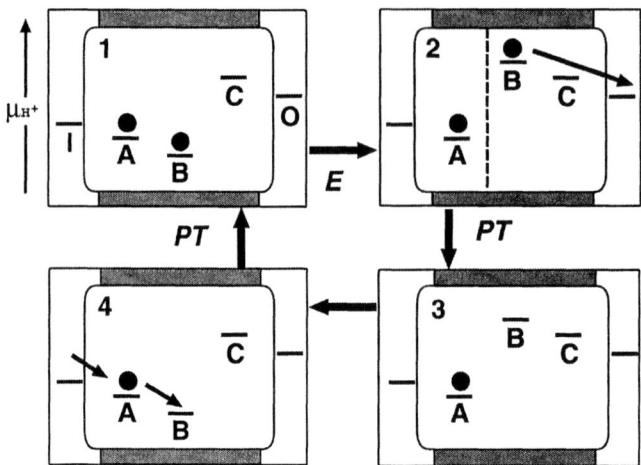

Figure 2. Schematic representation of the energetics of a proton-pumping membrane protein (see text). The vertical axis represents proton activity: the lower the group, the higher its proton affinity. In this cycle, a proton is pumped from I to O via three groups A, B, and C embedded in the protein. The protonation of these groups is indicated by large black dots. (1) resting state of the pump: A and B are protonated, C is deprotonated. (2) After energy uptake, B acquires high proton activity; the vertical dashed line stresses that its proton cannot leak back to I via A, but instead it is translocated to O via C. (3) After proton release, B relaxes back to its low-activity state. (4) B is reprotonated from I via A. The protonation state of proton-relaying groups A and C only changes transiently.

Group A, which has lower H^+ activity than I, is only transiently in its deprotonated state and inversely, C, whose proton activity is higher than O, is only protonated transiently. In Fig.2, the proton pumping cycle is depicted as arbitrarily starting from (1). In state (1) of the enzyme, the proton activity of group B is lower than those of A and I so that B is protonated. As a result of energy uptake (coupling), this proton becomes activated (state 2) and is subsequently released to the output channel via group C (state 3). After deprotonation (3), group B returns to its low proton affinity state (4), where it readily gets reprotonated from I via A, a step which is facilitated by the low proton affinity of A. This proton uptake completes the pumping cycle.

1.3 Thermodynamic and kinetic control

Consideration of relative proton affinities alone is not sufficient to explain the directionality of H^+ transport in proton pumps. For efficient proton pumping it is essential that the activated proton (state 2) cannot flow back to group A, which thermodynamically would be dictated by the fact that A has much higher proton affinity than C. To that effect, relative insulation to proton transport in state 2 is required either between B and A, or alternatively, between A and surface I. This requirement is often described in terms of an *alternating access model* [Jardetzky, 1966] and is now fairly well understood in bacteriorhodopsin [Lanyi, 1999]. Likewise, proton leakage from C to B must be prevented in step 4, where B has a much higher proton affinity than C. In the depicted scheme (Fig.2), the back flow of H^+ is hampered by the high proton activity of C relative to O, so that proton relay to B from A is rapid compared to alternative reprotonation from O via C.

Thus, both thermodynamic and kinetic control of proton transport must be achieved in proton pumps: thermodynamic, to overcome an electrochemical proton gradient; and kinetic, to prevent the back-flow of protons. For these reasons, hydrogen-bonded networks necessarily comprise regions where pumping and ratchet mechanisms operate. By contrast, the rest of the proton-relay pathway may function as "passive" plumbing, i.e., proton wires whose primary role is to transport protons in a timely fashion, consistently with the physiological turnover rate of the enzyme. For an efficient pump it is *a priori* preferable that the pumping (coupling) be localized, so as to minimize energy dissipation. In the idealized schemes of Fig.2, this is underlined by the proton activation of a single element, B, whereas the proton affinities of relay groups A and C located respectively in the uptake and the release pathways are depicted as essentially unchanged throughout the pumping cycle. By the same token, large portions of the proton path can be expected to be essentially devolved to plumbing.

Finally, the smaller the differences in proton affinity of the various groups in the uptake (release) pathway, the greater the efficiency of the pump. This requirement underlines the relevance of passive proton relay, as found in water-filled channels such as gramicidin, and provides a rationale for studying such relatively simple systems in order to better understand proton-relay in proton pumps.

1.4 How are those requirements fulfilled at the molecular level?

In order to understand the factors controlling plumbing and pumping, it is very important to explore the detailed molecular properties of proton wires in membrane proteins. In particular, it is essential to determine the subtle interplay between structural, equilibrium, and non-equilibrium properties in the molecular mechanism of proton transport in complex hydrogen-bonded networks.

This represents a formidable challenge, because all the following factors must in principle be characterized in detail throughout the pumping cycle: both (i) relevant electronic states of the protein and (ii) proton affinities of the various groups involved in the proton-shuttling pathway must be known in order to describe the thermodynamic driving forces for proton transport. Moreover, (iii) all the relevant conformational states of the protein (which are themselves coupled to the protonic and electronic states involved in the proton pumping mechanism), all the way down to the structure of hydrogen-bonded networks, and (iv) the dynamic transitions between the various intermediates in the proton pathway, including proton-transfer and reorganization steps, must be investigated for a complete description of the properties governing the directionality of the translocation.

The rest of this discussion deals with what may be expected to constitute the simpler portions of proton wires: those involved primarily in plumbing. Specifically, molecular mechanisms mediating H^+ translocation in water-filled channels are considered. Such wires constitute a pathway for the passive relay of H^+ in narrow transmembrane pores, and they provide models for the study of proton transport in water-filled channels embedded in complex energy-transducing proteins. Such models may in the future help in delineating which respective portions of the hydrogen-bonded networks in proton pumps are responsible for insulating, plumbing, gating, and pumping.

2 Proton-Conducting Channels

One thing that the energy-transducing proteins appear to have in common is the inclusion of water in pathways for proton translocation. Relay models of protons by buried water molecules have been substantiated in several systems of bioenergetic interest. For example, there are evidences for the involvement of buried water molecules in bacteriorhodopsin [Cao et al., 1991; Pebay-Peyroula et al., 1997]. In addition, proton-relay pathways involving chains of water molecules have been detected in high-resolution crystal structures of highly homologous bacterial photosynthetic reaction centers [Ermler et al., 1994;

Abresch et al., 1998] and of the lumen-side domain of cytochrome f [Martinez et al., 1996]. Finally, in cytochrome c oxidase, two independent theoretical studies have predicted the hydration of buried cavities implicated in the uptake of protons [Riistama et al., 1997; Hofacker and Schulten, 1998]. Although these water molecules were not resolved in published crystallographic structures of the *P. denitrificans* [Iwata et al., 1995] and bovine heart [Tsukihara et al., 1996] enzymes, many of them are well-defined in a new structure of the *Rb. sphaeroides* enzyme [M. Svensson-Ek, personal communication]. Understanding the molecular properties giving rise to proton transport in hydrogen-bonded networks containing water molecules is therefore an important step towards the elucidation of proton-pumping mechanisms.

In the past few years, theoretical approaches have provided detailed studies of the molecular mechanism for long-range proton transport in aqueous systems. While several of these studies have focussed on proton transport in bulk water [Tuckerman et al., 1995; Vuillemier and Borgis, 1998; Schmidt and Voth, 1999], others have been dedicated to hydrogen-bonded networks of water molecules in confined environments [Pomès and Roux, 1995, 1996a, 1996b, 1998; Pomès, 1999; Sagnella and Voth, 1996; Sagnella et al., 1996; Drukker et al., 1998; Decornez et al., 1998; Mei et al., 1998; Sadeghi and Cheng, 1999]. Because of their direct relevance to proton relay in membrane proteins, the latter systems are reviewed here. The simplest systems, nonpolar water-filled channels, are first considered, and a survey of studies of gramicidin and other membrane-spanning pores follows.

2.1 Model Non-polar Channels

This subsection reviews some of the results that have emerged from computer simulations of protonated chains of water molecules embedded in an idealized nonpolar (NP) cylinder and forming a linear hydrogen-bonded array. Because of the absence of hydrogen bond donors and acceptors on the channel wall, each water molecule can form at most two hydrogen bonds, one with each adjacent water in the single file arrangement. Detailed computational studies of such simple models of proton wires have led to an emerging mechanism for proton transport [Pomès and Roux, 1995, 1996a, 1998; Drukker et al., 1998; Decornez et al., 1998; Mei et al., 1998; Sadeghi and Cheng, 1999].

Although a number of methods and approximations have been used in these various studies, they are in general qualitative agreement with each other in their description of the proton-hopping mechanism. The protonated water wire is polarized by the presence of an excess proton. The polarization is strongest close to the excess charge. Consequently, the excess proton itself is often shared by two adjacent water molecules in a particularly strong and short (2.4 – 2.5 A) hydrogen bond. In such protonated water dimers, which are sometimes called Zundel cations, H^+ behaves as a quantum particle delocalized between the two O atoms.

The dynamic exchange of H^+ between adjacent pairs of O atoms involves structural fluctuations in the hydrogen-bonded chain that take place spontaneously at 300 K. Thermal motions of the heavy (O) atoms of the chain result in fluctuations of hydrogen-bond lengths. Occasionally, this triggers an exchange of the location of the shared proton, thereby giving rise to the unitary transport of the ionic defect in the chain. In chains of nine water molecules extending over 2 nm, such motions exhibit a strong semi-collective character, but they are not totally concerted [Pomès and Roux, 1996a]. In this mechanism, the translocation of protons across several water molecules may occur in picosecond [Pomès and Roux, 1996a] or even sub-picosecond time scales [Sadeghi and Cheng, 1999].

In simulations of a wire of nine water molecules flanked by water droplets mimicking the influence of bulk water at either end of a membrane pore, proton migration was found to be activation-less throughout the single-file chain [Pomès and Roux, 1998], a result also suggested by a study of a tetrameric chain with droplets [Mei et al., 1998]. Thus, thermal fluctuations of proton wires in model hydrophobic channels are sufficient to drive the complete translocation within a few ps, despite the absence of an external field acting as driving force.

The turn step of the Grotthuss mechanism was studied in an unprotonated chain of nine water molecules embedded in a NP channel [Pomès and Roux, 1998]. In the absence of an excess proton, the chain is preferentially polarized: all the successive OH bonds engaged in water-water hydrogen bonds point in the same direction, in agreement with the idealized picture of the Grotthuss mechanism (Fig.1). The exchange between these two conformations (reorientation of the wire) involves a substantial activation energy barrier (7 to 8 kcal/mol), in striking contrast with the activation-less PMF profile for proton hopping. The reorientation process involves a succession of transient configurations, each of which comprises a single bonding defect. The water molecules reorient sequentially, which causes the defect to propagate along the chain.

2.2 Gramicidin

The rapid equilibration of H^+ across biological membranes is accomplished by proton conducting channels [DeCoursey and Cherny, 2000]. Channels are membrane-spanning proteins forming water-filled pores, of which Gramicidin A (GA) is the best-understood example [Tian and Cross, 1999; Wallace, 1998]. In its active form, GA assembles as a dimer in lipid bilayers. Its alternating L and D amino acids fold into a right-handed beta-helix structure, which exposes its hydrophobic side chains to the surrounding lipid bilayer, while the peptide backbone lines the interior of a cylindrical pore 0.4 nm in diameter. This hydrophilic pore, which is just wide enough to accommodate a single file of 8 to 10 water molecules, mediates the translocation of monovalent cations such as H^+, Cs^+, Na^+, and K^+ [Finkelstein, 1987]. Strong evidence for a Grotthuss mechanism for proton conduction in GA is provided by the lack of osmotic streaming potentials during H^+ permeation [Levitt et al., 1978]. Contrary to the permeation of other ions, which displaces the single-file column of water out of the channel, the hopping of protons does not involve net diffusion of water molecules.

The solvation of a hydronium ion [Sagnella and Voth, 1996] and the translocation of an excess proton in GA [Pomès and Roux, 1996b; Sagnella et al., 1996] were studied with computer simulations. These studies underline the role of the backbone carbonyl O atoms lining the pore of the channel in the coordination and the dynamics of protonated water chains. Compared to that of simple water wires embedded in NP channels, in GA the hydrogen-bonded network gains dimensionality: in addition to the possibility of forming up to two hydrogen bonds with adjacent water molecules in the single file, each water molecule can also donate hydrogen atoms to carbonyl O atoms lining the pore. This feature, which results in the optimal coordination of both OH_3^+ and $O_2H_5^+$ ions [Pomès, 1999], also has important consequences for both structural and dynamic properties of the water wire, as evidenced in relatively long proton-hopping simulations [Pomès and Roux, 1996b].

In that study, the analysis of the structure and fluctuations of the hydrogen-bonded network formed by the single-file water molecules led to the distinction of two regions. Near the excess charge, the water molecules are strongly polarized and form a well-connected chain of water molecules extending over much of the pore's length. In this polarized cluster of up to six or seven water molecules, hopping of H^+ is similar to that observed in NP channels (see above), and takes place spontaneously with thermal fluctuations in the picosecond time-range [Pomès and Roux, 1996b].

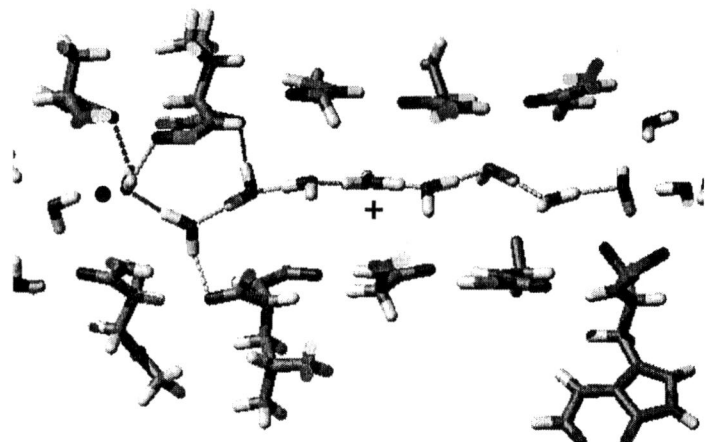

Figure 3. Water wire in the GA channel in the presence of H^+. A slice of the channel is shown, together with a few hydrogen bonds. The location of the ionic defect is indicated with a + sign. Proton hopping takes place spontaneously along the continuous hydrogen-bonded water chain, in which each water molecule donates one H to a backbone carbonyl O atom, and one to another water. A water molecule donating both H atoms to carbonyl groups lining the cylindrical pore creates a bonding defect, highlighted by a dot.

Further away from the average location of the excess proton, bonding defects occur whenever a single-file water molecule donates both of its H nuclei to the pore. As depicted in Fig.3, these bonding defects confine the extension of the polarized water chain and block further hopping of H^+ [Pomès and Roux, 1996b]. Migration of such defects (ie, the turn step of the Grotthuss mechanism) necessitates the reorientation of water molecules. In the interior of the GA channel, water reorientation is slow compared to proton hopping in the polarized cluster: while hopping takes place in the ps time range, the lifetime of bonding defects could be of the order of 0.1 ns or longer. Thus, the turn step of the Grotthuss mechanism appears to constitute the rate-limiting step for the fast translocation of protons in the single-file region.

2.3 Leakage vs. Conduction Mechanisms

The results summarized above suggest important mechanistic differences in the long-range relay mechanism of protons by water wires embedded in non-polar (NP) and polar (GA) channels, respectively. Whereas GA constitutes an example of of fast proton equilibration by water-filled pores, NP channels may be relevant to transient hydrogen-bonded chain models proposed for proton leakage in lipid membranes. Leakage of protons occurs even in pure lipid membranes, but it is a very infrequent phenomenon. To explain leakage, it was proposed that water molecules present in the non-polar core of the lipid membrane assemble transiently to form hydrogen-bonded chains [Nagle, 1987; Deamer and Nichols, 1989]. Experimental support for H^+ leakage by transient hydrogen-bonded water chains was discussed for a membrane thickness of 2.2 to 3 nm [Paula et al., 1996]. The thermodynamic basis for the formation of an unprotonated single-file of water molecules in a pure lipid membrane was examined with molecular dynamics simulations of an explicit lipid bilayer [Marrink et al., 1996]. It was concluded that in order to explain experimental measurements of leakage, each transient water wire would have to transport one proton within its lifetime. Simulations of proton hopping suggest that such a requirement can be met in single-file water chains: in model NP channels, unprotonated water chains are preferably polarized or primed for proton transport, and complete translocation of an excess H^+ is very fast [Pomès and Roux, 1998]. Specifically, the hop of a proton along a water chain of nine water molecules was seen to occur in the order of a picosecond [Pomès and Roux, 1996b, 1998]. However, the strong activation required to achieve the reorientation of the chain also suggests that the turn process does not have sufficient time to occur during the lifetime of a transient water wire in the bilayer. Taken together, these results suggest that once formed, the wire could relay the hopping of up to one proton before the hydrogen-bonded chain breaks up into its disordered state, with water molecules dispersed throughout the bilayer [Pomès and Roux, 1998] (see Fig.4). In such a mechanism, a transient wire would perform only the hopping half of the Grotthuss mechanism before reverting to its dislocated state.

In contrast to the transient hydrogen-bonded chain mechanism proposed for "leakage," the mechanism leading to the equilibration of protons across the membrane utilizes comparatively long-lived water chains that are thermodynamically stabilized by a polar, hydrogen-bonding environment such as that provided by the GA pore. The lifetime of the GA dimer in lipid membranes is of the order of 100 ms [Andersen, 1984], and spectroscopic studies indicate that the pore is permanently occupied by water molecules [Finkelstein, 1987]. In that environment, proton conduction proceeds at a rate of up to 10^9 s^{-1} [Cukierman, 1999]. These considerations lead to the distinction of two different mechanisms, which may be called respectively *leakage* and *conduction*, for the relay of protons in polar and non-polar cavities. The basic difference between these two limiting mechanisms is the physical origin of the rate-limiting step: in the case of conduction, it is the Grotthuss mechanism itself, whereas in the case of leakage, it is the nucleation of water molecules in a non-polar environment.

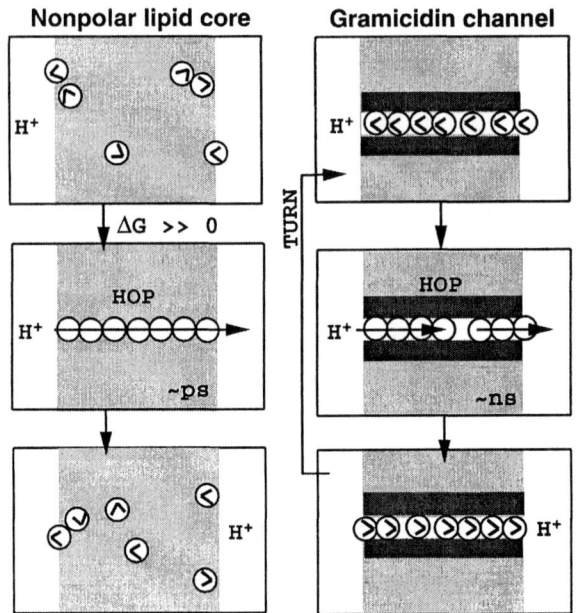

Figure 4. Leakage vs. conduction models for the permeation of protons. *Left: transient water wires form very infrequently in the nonpolar environment provided by lipid bilayers, but when they do, they could translocate just one proton very rapidly before breaking up; only the hop step of Grotthuss takes place. Right: in polar channels such as GA, water wires are much more long-lived, which is consistent with their rapid relay of proton via a* complete *Grotthuss mechanism involving both hop and turn steps ($10^7 \, s^{-1}$).*

By analogy, it is conceivable that energy-transducing proteins make use of both leakage- and conduction-type mechanisms, as was proposed based on a comparison of proton-translocation pathways in bacteriorhodopsin and in cytochrome c oxidase [Wikström, 1998]. Remarkably, the proton uptake pathways in the interior of both enzymes include relatively non-polar regions [see Wikström, 1998, and references therein]. Accordingly, the highest-resolution crystallographic structures obtained to date [Luecke et al., 1999; M. Svensson-Ek, personal communication] reveal incomplete connectivity of proton-relay groups due to partial hydration in the uptake pathways. However, while these observations support the presence of at least partly-disordered hydrogen-bonded water chains for proton uptake, further studies are required to establish whether transient hydrogen-bonded chains play a role in the kinetic control of directional proton relay in proton pumps, as evoked above.

3 Conclusions

The computational study of proton transport in pure water wires has uncovered important aspects of the hop and turn (Grotthuss) mechanism at the molecular level. In particular, the respective roles of thermal fluctuations, of nuclear quantum effects, and of the modulation of the wire's properties by a protein matrix have been characterized in simple, yet biologically-relevant systems. A significant result is that proton translocation can be dynamically limited by the intrinsic properties of hydrogen-bonded networks. Thus, structural fluctuations of the hydrogen-bonded network in these simple water wires, as in bulk water, appear to govern the molecular mechanism for long-range proton transport. Both in NP and GA channels, computational studies suggest that it is easier to translocate an ionic defect than a bonding defect. These results provide an impetus for the detailed study of hydrogen-bond network structure and dynamics in other membrane proteins---particularly in energy-transducing proteins, where large tracts of the proton-relay pathways are occupied by water molecules and appear to constitute much of the plumbing.

Acknowledgments

This work was supported by the US Department of Energy through the Los Alamos National Laboratory LDRD grant for Bioremediation and by start up funds from the Hospital for Sick Children.

Note added in proof

Since the original submission of this contribution, new computational studies of proton mobility in NP channels [Brewer, M.L., Schmitt, U.W., and Voth, G.A. (2001) Biophys. J. 80, 1691-1702] and of the hop-and-turn mechanism for proton conduction in GA [Pomès, R., and Roux, B. (2002) Biophys. J. 82, 2304-2316] have been published.

References

Abresch, E.C., Paddock, M.L., Stowell, M.H.B., McPhillips, T.M., Axelrod, H.L., Soltis, S.M., Rees, D.C., Okamura, M.K., and Feher, G. (1998) Photosynth. Res. 55, 119-124.
Agmon, N. (1995) Chem. Phys. Lett. 244, 456-462.
Andersen, O.S. (1984) Ann. Rev. Physiol. 46, 531-548.

Brunger, A., Schulten, Z., and Schulten, K. (1983) Zeit. Phys. Chem. 136, 1-63.
Cao, Y., Varo, G., Chang, M., Ni, B., Needleman, R., and Lanyi, J.K. (1991) Biochemistry 30, 10972-10979.
Cukierman, S. (1999) Isr. J. Chem.39, 409-418.
Deamer, D.W., and Nichols, J.W. (1989) J. Membr. Biol. 107, 91-103.
Decornez, H., Drukker, K., and Hammes-Schiffer, S. (1999) J. Phgys. Chem. A 103, 2891-2898.
DeCoursey, T.E., and Cherny, V.V. (2000) Biopchim. Biophys. Acta 1458, 104-119.
De Grotthuss, C.J.T. (1806) Ann. Chim. 58, 54-74.
Drukker, K., de Leeuw, S.W., and Hammes-Schiffer, S. (1998) J. Chem. Phys. 108, 6799-6808.
Ermler, U., Fritzsch, G., Buchanan, S.K., and Michel, H. (1994) Structure 2, 925-936.
Finkelstein, A. (1987) Water movement through lipid bilayers, pores, and plasma membranes. John Wiley & Sons, New York.
Hofacker, I., and Schulten, K. (1998) Proteins Struct. Funct. Gen. 30, 100-107.
Iwata, S., Ostermeier, C., Ludwig, B., and Michel, H. (1995) Nature 376:660-669.
Jardetzky, O. (1966) Nature 211, 969-970.
Knapp, E.W., Schulten, K., and Schulten, Z. (1980) Chem. Phys. 46, 215-229.
Lanyi, J.K. (1999) FEBS Lett. 464, 103-107.
Levitt, D.G., Elias, S.R., and Hautman, J.M. (1978) Biochim. Biophys. Acta 512, 436-451.
Luecke, H., Schobert, B., Richter, H.T., Cartailler, J.P., and Lanyi, J.K. (1999) Science 286, 255-261.
Marrink, S.J., Jahnig, F. and Berendsen, H.J.C (1996) Biophys. J. 71, 632-647.
Martinez, S.E., Huang, D., Ponomarev, M., Cramer, W.A., and Smith, J.L. (1996) Protein Science 5, 1081-1092.
Mitchell, P. (1961) Nature 191, 144-148.
Mei, H.S., Tuckerman, M.E., Sagnella, D.E. and Klein, M.L. (1998) J. Phys. Chem B 102, 10446-10458.
Nagle, J.F. and Morowitz, H.J. (1978) Proc. Natl. Acad. Sci. USA 75, 298-302.
Nagle, J.F. (1987) J. Bioenerg. Biomembr. 19, 413-426.
Paula, S., Volkov, G., van Hoek, N., Haines, T.H., and Deamer, D.W. (1996) Biophys. J. 70, 339-348.
Onsager, L. (1967) in "Neurosciences," ed. F.O. Schmitt, Rockefeller University Press, New York, 75-79.
Pebay-Peyroula, E., Rummel, G., Rosenbusch, J.P., and Landau, E.M. (1997) Science 277, 1676-1681.
Pomès, R. (1999) Isr. J. Chem. 39, 387-395.
Pomès, R. and Roux, B. (1995) Chem. Phys. Lett. 234, 416-424.

Pomès, R. and Roux, B. (1996a) J. Phys. Chem. 100, 2519-2527.
Pomès, R. and Roux, B. (1996b) Biophys. J. 71, 19-39.
Pomès, R. and Roux, B. (1998) Biophys. J. 75, 33-40.
Riistama, S., Hummer, G., Puustinen, A., Dyer, R.B., Woodruff, W.H., and Wikstrom, M. (1997) FEBS Lett. 414, 275-280.
Sadeghi, R.R., and Cheng, H.P. (1999) J. Chem. Phys. 111, 2086-2094.
Sagnella, D.E., Laasonen, K. and Klein, M.L. (1996) Biophys. J. 71, 1172-1178.
Sagnella, D.E. and Voth, G.A. (1996) Biophys. J. 70, 2043-2051.
Saraste, M. (1999) Science 283, 1488-1493.
Schmidt, U.W., and Voth, G.A. (1999) J. Chem. Phys. 111, 9361-9381.
Tian, F. and Cross, T.A. (1999) J. Mol. Biol. 285, 1993-2003.
Tsukihara, T., Aoyama, H., Yamashita, E., Tomizaki,T., Yamaguchi, H., Shinzawa-Itoh, K., Nakashima, R., Yaono, R. and Yoshikawa, S. (1996) Science 272, 1136-1144.
Tuckerman, M., Laasonen, K., Sprik, M. and Parrinello, M. (1995) J. Phys. Chem. 99, 5749-5752.
Vuilleumier, R. and Borgis, D. (1998) J. Phys. Chem. B. 102, 4261-4264.
Wallace, B.A. (1998) J. Struct. Biol. 121, 123-141.
Warshel, A. (1979) Photochem. Photobiol. 30, 285-290.
Wikström, M. (1997) Nature 266, 271-273.
Wikström, M. (1998) Curr. Opin. Struct. Biol. 8, 480-488.

Chapter 11

Computer Simulation of Energy-Transducing Proteins and Peptide:Membrane Interactions

Dan Mihailescu[1,2], G. Matthias Ullmann[2], and Jeremy C. Smith[2,*]

[1]Faculty of Biology, University of Bucharest, Spl. Independentei, 91–95, 76201, Bucharest, Romania
[2]Lehrstuhl für Biocomputing, IWR, Universität Heidelberg, Im Neuenheimer Feld 368, D–69120 Heidelberg, Germany

We summarise recent work on computer modelling and simulation of proteins involved in bioenergetic processes and in peptide-membrane interactions. Homology modelling, electrostatic calculations and conformational analysis of a photosynthetic reaction centre protein are described. Bacteriorhodopsin, a light-driven proton pump protein is examined from several aspects, including its hydration and conformational thermodynamics. Finally, we present results on lipid perturbation on interaction with a cyclic decapeptide antibiotic, gramicidin S.

Introduction

There is much interest in understanding the structure and function of proteins and peptides in interaction with lipid bilayers. This is particularly so in bioenergetics where, for example, several important energy transducing proteins are membrane spanning. We summarise here recent results aimed at characterising two aspects of these processes:
- the structure and function of photosynthetic proteins (photosynthetic reaction centres and bacteriorhodopsin).
- recent results on peptide-membrane interactions.

Photosynthetic Reaction Centre

The photosynthetic reaction centres (RCs) are transmembrane protein-pigment complexes that perform light-induced charge separation during the primary steps of photosynthesis. RCs from purple bacteria consist of three protein subunits, L, M and H, and bind four bacteriochlorophylls, two bacteriopheophytins, two quinones, one non-haem iron and one carotenoid. The elucidation at atomic resolution of the three-dimensional structures of the bacterial RCs from *Rhodopseudomonas (Rps.) viridis* (1) and *Rhodobacter (Rb.) sphaeroides* (2-4) has provided impetus for theoretical and experimental work on the mechanism of primary charge separation in the RCs. The structures revealed that the cofactors are bound at the interface between the L and M subunits and are organised around a pseudo C2 symmetry axis. However, the structural symmetry does not result in functional symmetry as the electron transfer proceeds only along the L branch (5).

Much of the experimental structure-function work on RCs uses mutagenesis. Site-directed mutagenesis has been most developed for the RC from *Rhodobacter capsulatus*. Using this technique the *Rb. capsulatus* RC has been partially symmetrized (6). One of the partially symmetrized mutants, named D_{LL}, has turned out to be of particular interest (7). In the D_{LL} mutant the M subunit D helix is replaced by its L subunit counterpart. The D helix was chosen because it contacts all the pigments on a given branch. An unexpected outcome of the partial symmetrization was that the D_{LL} mutant lacks the L-branch bacteriopheophytin (BPh_L), which plays an essential role as a transient electron carrier in native RCs. Four revertants that do bind BPh_L at least partially, and can grow photosynthetically, were isolated from the D_{LL} mutant (8,9). The fact that the D_{LL} mutant does not bind BPh_L renders it particularly useful for experiments on primary electronic processes occurring before electron transfer to BPh_L. Indeed, the D_{LL} mutant was used to demonstrate the presence of coherent nuclear motion coupled to stimulated emission from the primary donor (7).

Information on the structures of the D_{LL} protein and the revertants is necessary to provide a basis for descriptions of the electronic excitation and BPh_L binding properties of these systems. To obtain this a model structure for the native RC of *Rb. capsulatus* was derived by combining structural information from the X-ray crystallographic analysis of the closely-related RC of *Rps. viridis* with molecular mechanics energy calculations (10). In the D_{LL} mutant the orientation of the five remaining pigments with respect to the C2 axis is essentially the same as in the wild type, suggesting that the structures of these proteins are similar (8,9). Starting from the wild-type homology model conformational energy calculations on the D_{LL} mutant and several BPh_L-binding revertants were subsequently performed (11). They provide an explanation for the relative BPh_L binding properties of the proteins in terms of interactions involving two residues in the binding pocket, these being a tryptophan and a methionine in the D_{LL} protein.

The absorption of light by the RC leads to an electron transfer. The final electron acceptor is the ubiquinone Q_B, which receives subsequently two electrons. The reduction of Q_B is coupled to protonation state changes of Q_B itself and of the protein environment. The coupling of the protonation and reduction steps can be investigated by continuum electrostatic calculations (12-14). One application of this method to the RC indicates that the proton uptake by the RC is coupled more strongly to changes of the redox state of Q_B than to changes of it protonation state (15). Thus, the proton uptake by the RC occurs predominantly before the protonation of Q_B. The reduction of Q_B^- to the doubly negative state Q_B^{2-} is in the RC energetically even more unfavorable than in solution. Therefore, the second electron transfer to Q_B occurs after Q_B has received its first proton.

The reduction of Q_B is coupled to a conformational rearrangement. The crystal structures of the dark-adapted and the light-exposed RC from *Rb. sphaeroides* were solved and the conformational changes were characterized structurally (16). In the two structures, the Q_B was found in two different positions, proximal or distal to the non-haem iron. Because Q_B was found mainly in the distal position in the dark and only in the proximal position under illumination, the two positions have been attributed mostly to the oxidized and the reduced forms of Q_B, respectively. According to electrostatic calculations, the reaction energy of the electron transfer to Q_B is energetically uphill at pH 7 for the dark-adapted structure when Q_B is in the distal position and downhill for the light-exposed structure when Q_B is in the proximal position (17,18).

An analysis of the proton uptake upon reduction of Q_B as a function of pH (19) in combination with continuum electrostatic calculations suggests, that the conformational equilibrium depends both on the redox state of Q_B and on the pH of the surrounding medium (20). Q_B occupies only the distal position below pH 6.5 and only the proximal position above pH 9.0 in both oxidation states. Between these pH values both positions are partially occupied. The reduced Q_B has a higher occupancy in the proximal position than the oxidized Q_B.

A study using molecular dynamics simulation has shown that the proximal position of Q_B^- is more stable when the two adjacent residues L212-Glu and L213-Asp are both protonated (21). However, electrostatic calculations show that the position of Q_B does not depend only on the protonation state of L212 and L213, but also on the protonation state of other residues that trigger the conformational transition between RC^{prox} and RC^{dist}. This finding is also supported by a more recent molecular dynamics simulation study (22).

Molecular dynamics simulations of the RC of *Rps. viridis* have provided additional evidence supporting the movement of Q_B between the distal site and the proximal site (23). This work showed that the equilibrium between the two binding sites is not displaced by the reduction of Q_B to the semiquinone, by the preceding reduction of the primary quinone Q_A and by accompanying protonation changes in the protein.

Bacteriorhodopsin

The light-driven proton pump, bacteriorhodopsin (bR), is found in the purple membrane of *Halobacterium halobium* (24-26).

The experimental structure of bR determined at atomic resolution from cryoelectron microscopy and X-ray crystallography revealed a channel containing the Schiff base of the retinal chromophore (27, 28). Site-directed mutagenesis and vibrational spectroscopy experiments have enabled the identification of polar residues in the channel involved in the proton transfer pathway (29-32). Recent work on bacteriorhodopsin has concentrated on hydration and conformational thermodynamics.

Buried cavities in membrane proteins can be occupied by water molecules, which can play important roles in stabilizing protein structure and in functional mechanisms. Whether a given cavity is occupied by water depends on the environment it provides. A qualitative appreciation of this can sometimes be gained by examining the size and shape of the cavity and the hydrophobic or hydrophilic nature of the side-chains lining it. However, a more rigorous analysis requires the determination of thermodynamic quantities, in particular the free energy of transfer of a water molecule from bulk solvent to the buried site considered. A theoretical determination of this quantity is therefore of basic interest in understanding cavity hydration. Moreover, theoretical analysis may be of use in analyzing the likelihood of occupancy of sites not amenable to experimental determination. For example, this problem can arise in crystallographic analyses of water molecules with partial occupancies and/or high thermal fluctuations or at a resolution insufficient for their characterisation.

In principle the free energy of transfer can be determined using molecular dynamics simulations. However, as residence times of buried water molecules can be on the ms timescale or longer, indicating the presence of significant energy barriers to their migration, water molecules are not expected to partition into the interior of proteins according to thermodynamic equilibrium in standard ps-ns timescale molecular dynamics calculations. Consequently, specialised molecular dynamics techniques must be employed.

We have developed a rigorous formulation for calculating the thermodynamic stability of water molecules inside a protein cavity (33).

Mathematical expressions, suitable for evaluation from computer simulations, were derived for the binding constant and the probability of finding any number of water molecules in isolated cavities. The formalism is general and can be applied to investigate the binding constant and the probability of occupancy of any molecule in a specific site in a macromolecule.

Several lines of evidence indicate that water molecules may also be present in the bacteriorhodopsin channel and may play an important functional role. While the cryoelectron microscopy structure lacks sufficient resolution to locate water molecules, a contrast variation neutron diffraction study has indicated that there are about four water molecules present in the neighborhood of the Schiff base (34). Spectroscopic experiments suggest that one or more water molecules are directly hydrogen-bonded to the Schiff base (35-41).

In initial theory papers, ab initio quantum chemical calculations were performed on Schiff base-water complexes to investigate possible direct hydrogen-bonding interactions (42,43). Two minimum-energy interaction sites were found; one involves a hydrogen bond with the Schiff base NH group, which in light-adapted bR is in a polar environment on the extracellular side of the Schiff base, and the other involves a CH....O hydrogen bond on the cytoplasmic side. These were the only two significant minima on the potential surface.

Other water molecules, not directly associated with the Schiff base, may also be involved in the proton transfer pathway. The reprotonation step at the end of the photocycle involves the transfer of a proton from Asp 96, a residue near the cytoplasmic surface, to the retinal Schiff base (44-48). This requires the translocation of a proton over a distance of 10 to 12 Å through a narrow region of the channel lined with nonpolar residues. The question arises as to how the proton is transferred across this region. One possibility is that a number of water molecules could form a proton transfer chain from Asp 96 to the Schiff base. While the chain of water molecules could be stabilized by making hydrogen bonds to each other, and perhaps to Thr 46 and Thr 89, preliminary analysis of the bR structure suggests that it would be primarily located in a nonpolar cavity. This raises important questions concerning the thermodynamic stability of water molecules in such an environment. To investigate this possibility we applied the free energy methodology to investigate the thermodynamic stability and probability of occupancy of water molecules in specific sites in the channel of bR (33). The results suggest that the transfer of water molecules from bulk water to bR so as to form an intact hydrogen-bonded column between the proton donor, Asp 96, and the Schiff base, is thermodynamically permitted.

In subsequent work the phenomenon of dark adaptation was examined. After several minutes in the dark bacteriorhodopsin contains a mixture of two isomers of retinal, one which is *all–trans* and the other *(13,15)-cis*. The relative populations of these two forms correspond to a free energy difference of about kT with the *cis* form being the more populated. In recent work adiabatic potential energy calculations and umbrella sampling free energy calculations were performed (using the WHAM procedure) to investigate factors influencing this equilibrium (49,50). Agreement with experiment to within kT was found, and several important factors influencing the equilibrium, including the water molecules, were determined. In a recent paper, the bacteriorhodopsin dark adaptation procedure was used as a test case in examination of efficient methods for calculating free energy profiles along biologically interesting reaction coordinates (51). Finally, transitions from harmonic to anharmonic dynamics in bacteriorhodopsin, which are correlated with function, have been analyzed using molecular dynamics (52).

Peptide-Membrane Interactions

The examination of peptide membrane interactions requires characterisation of the position, orientation, structure and dynamics of the peptide in the lipid bilayer as well as its effects on surrounding lipids. While molecular dynamics simulation can in principle furnish complete structural and dynamical information, considerable obstacles exist to obtaining accurate results, due partly to inexact force fields and other approximations in simulation methodology, and partly to the relaxation times of some important dynamical phenomena being of the order of or longer than those presently accessible to MD.

In recent work gramicidin S, [*cyclo*-(Leu-D-Phe-Pro-Val-Orn)$_2$, (GS)] a cyclic decapeptide (Figure. 1) that is particularly suitable for pursuing peptide:membrane MD studies, was examined in interaction with a DMPC membrane. The sequence and structure of the peptide are relatively simple. NMR, X-ray and MD studies indicate that the backbone adopts an antiparallel β-sheet with two Type II' β-turns in various solutions of different polarity and in the crystalline form (53-55). One consequence of this is that the GS structure is amphipathic, with the hydrophobic side chains on one side of the molecule and the hydrophilic ones on the other, and this provides a logical geometry for interaction with a lipid membrane, with the polar side of GS at the lipid:water interface and the nonpolar side interacting with the lipid tails. Considerable experimental evidence exists that this is indeed the case and a variety of biophysical studies on the interaction of GS with model lipid membranes have confirmed that it interacts primarily with the headgroup and interfacial polar/apolar regions (56-60). Also of interest is that GS has antibiotic action *via* membrane ion permeability change (61,62).

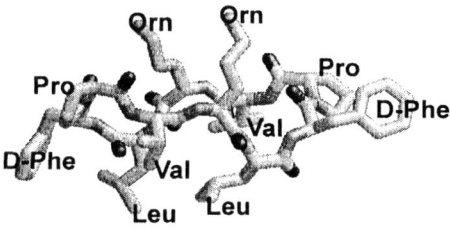

Figure 1. Gramicidin S

Although a complete description of GS ion permeability is probably beyond the capability of present-day simulation methods, information on how GS interacts with and perturbs lipid bilayers under simple controlled conditions may be obtained in this way. In initial simulation work we performed a 5 ns MD calculation of GS in DMSO solution, enabling detailed comparison with NMR results in the same solvent (55). Here we report on preliminary results on the interaction of GS with a hydrated DMPC membrane bilayer.

Two MD simulations were performed, one of GS in interaction with a hydrated DMPC bilayer and a 'control' simulation of a hydrated DMPC bilayer in the absence of GS. The control simulation was used to check the validity of the simulation method with respect to spectroscopic and diffraction data, and to use for comparison of the lipid molecule structure and dynamics in the presence and absence of GS. Detailed results will be described elsewhere.

During NVE equilibration GS diffused into the bilayer, with its centre of mass 3-4 Å deeper into the hydrophobic core than at the beginning of the equilibration. After about 1 ns of NVE equilibration the GS molecule regained its initial membrane/water interfacial position with the backbone and membrane planes parallel to each other. During the 3.0 ns NPT production dynamics the average position of GS relative to the membrane remained practically unchanged.

Nine lipids from the GS-containing layer have average interaction energies greater than k_BT, and in what follows these are considered as "bound" to the GS. The remaining eight GS-layer lipids are "free" as are the 21 from the non-GS layer. It is of interest to examine whether the ordering of the bound lipids is different from that of the free ones. A suitable quantity with which to examine this is the order parameter of the carbon-deuterium bond, S_{CD}, defined by $S_{CD} = <1/2\ (3\ \cos^2 \theta\ (t)\ -1)>$ where $\theta(t)$ is the angle between the carbon-deuterium bond vector and the bilayer normal. "<>" means 'time average'. For both kinds of lipids, an ensemble average and a time average for the two lipid tails was calculated to obtain the molecular order parameter plots ($-2S_{CD}$) in Figure 2. The figure also shows experimental data obtained using NMR quadrupole splitting experiments on fully-deuterated multilamellar dispersions of DMPC doped with GS at a molar ratio of 1:5.5 at 305K. (57). The results show that the free lipids are more ordered than in the control simulation. The bound lipid order parameters are close to the experimental values of Zidovetzki et al., 1988. Clearly, the lipids interacting with GS are more disordered.

Conclusions.

We have illustrated here how modelling and simulation can be applied to the understanding of membrane protein structure and function and the effect of peptides on lipid ordering. Evidently these calculations are but a small start towards understanding membrane protein structure and function at atomic detail, and the range of application is strongly limited by the availability of complementary experimental structural information. However, as described here, experiment has already furnished information of sufficient quality that a variety of atomic-detail computational techniques can usefully be applied. Future computer simulation research aimed at understanding functional photocycles,

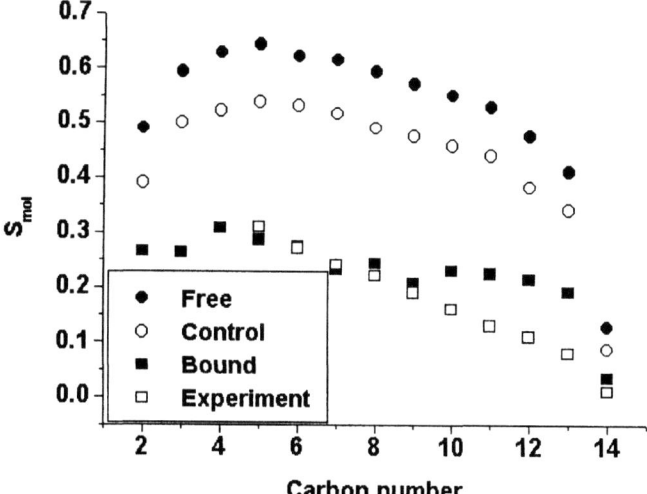

Figure 2. Molecular order parameter ($S_{mol}=-2S_{CD}$) averaged between the chains sn-1 and sn-2 of DMPC: "free" lipids (●); control simulation (○); "bound" lipids (■) and derived from NMR quadrupole splitting experiments on fully-deuterated multilamellar dispersions of DMPC doped with GS (□) (40)

electronic properties and protein:membrane interactions promises to reveal information of exceptional fundamental interest in structural cell biology.

References

1. Deisenhofer, J.; Epp, O.; Miki, K.; Huber, R.; Michel, H. *Nature* **1985**, 318, 618.
2. Allen, J.P.; Feher, G.; Yeates, T.O.; Komiya, H.; Rees, D.C. *Proc. Natl. Sci. USA* **1987**, 84, 6162.
3. Chang, C.H.; El-Kabbani, O.; Tiede, D.; Norris, J.; Schiffer, M. *Biochemistry* **1991**, 30, 5352.
4. Ermler, U.; Fritzsch, G.; Buchanan, S.K.; Michel, H. *Structure* **1994**, 2, 925.
5. Michel-Beyerle, M.E.; Plato, M.; Deisenhofer, J.; Michel, H.; Bixon, M.; Jortner, J. *Biochim. Biophys. Acta* **1988**, 932, 52.
6. Coleman, W.J.; Youvan, D.C. *Annu. Rev. Biophys. Biophys. Chem.* **1990**, 19, 333.
7. Vos, M.H.; Rappaport, F.; Lambry, J.C.; Breton, J.; Martin, J.L. *Nature* **1993**, 363, 320.
8. Robles, S.J.; Breton, J.; Youvan, D.C. *Science* **1990**, 248, 1402.
9. Robles, S.J.; Breton, J.; Youvan, D.C. *Reaction Centers of Photosynthetic Bacteria.* Michel-Beyerle, M.E., Eds.; Springer Verlag, Berlin, Germany, **1990**, pp. 283.
10. Foloppe, N.; Ferrand, M.; Breton, J.; Smith, J.C. *Proteins: Struct. Funct. Genetics* **1995** 22(3) 226.
11. Foloppe, N.; Breton, J.; Smith, J.C. *Chem. Phys. Lett.* **1995**, 242, 238.
12. Ullmann, G.M.; Knapp, E.W. *Eur. Biophys. J.* **1999**, 28, 533.
13. Ullmann, G.M. *J. Phys. Chem B* **2000**, 104, 6293.
14. Gunner, M.R.; Alexov, E. *Biochim Biophys Acta* **2000**, 1458, 63.
15. Rabenstein, B.; Ullmann, G.M.; Knapp, E.W. *Biochemistry* **1998**, 37, 2488.
16. Stowell, M.H.; McPhillips, T.M.; Rees, D.C.; Soltis, S.M.; Abresch, E.; Feher, G. *Science* **1997**, 276, 812.
17. Rabenstein, B., Ullmann, G.M.; Knapp, E.W. *Biochemistry* **2000**, 39, 10487.
18. Alexov, E.G.; Gunner, M.R. *Biochemistry* **1999**, 38, 8253.
19. Tandori, J., J. Miksovska, M. Valerio-Lepiniec, M. Schiffer, P. Maroti, D.K. Hanson, and Sebban, P. *Photochem. Photobiol.* **2002**, 75, 126
20. Taly, A.; Sebban, P.; Smith, J.C.; Ullmann, G.M. *Biophys. J.* **2003**, 84, in Press.
21. Grafton, A.K.; Wheeler, R.A.. *J. Phys. Chem. B* **1999**, 103, 5380.
22. Walden, S.E.; Wheeler, R.A. *J. Phys. Chem. B* **2002**, 106: 3001-3006.
23. Zachariae, U.; Lancaster, C.R. *Biochim Biophys Acta* **2001**, 1505:280.
24. Oesterhelt, D.; Stoeckenius, W. *Nature (London), New Biol* **1971**, 233, 149.

25. Oesterhelt, D.; Stoeckenius, W. *Proc. Natl. Acad. Sci. USA.* **1973**, 70, 2853.
26. Haupts, U.; Tittor, J.; Oesterhelt, D. *Ann. Rev. Biophys. Struct. Biol.* **1999**, 28, 367.
27. Henderson, R.; Baldwin, J.M.; Ceska, T.A.; Zemlin, F.; Beckmann, E.; Downing, K.H. *J. Mol. Biol.* **1990**, 213, 899-929.
28. Luecke, H.; Schobert, B.; Richter, H.-T.; Cartailler, J.-P.; Lanyi, J.K. *J. Mol. Biol.* **1999**, 291, 899.
29. Mogi, T.; Stern, L.J.; Hackett, N.R.; Khorana, H.G. *Proc. Natl. Acad. Sci. USA* **1987**, 85, 5595-5599.
30. Mogi, T.; Stern, L.J.; Marti, T.; Chao B.H.; Khorana, H.G. *Proc. Natl. Acad. Sci. USA* **1988**, 84, 5595-5599.
31. Stern, L.J.; Khorana, H.G. *J. Biol. Chem* **1989**, 264, 14202-14208.
32. Marti, T.; Otto, H.; Mogi, T.; Rosselet, S.J.; Heyn, M.P.; Khorana, H.G. *J. Biol. Chem* **1991**, 266, 6919-6927.
33. Roux, B.; Nina, M.; Pomes, R.; Smith, J.C. *Biophys. J.* **1996**, 71 670-681.
34. Papadopoulos, G.; Dencher, N.; Zaccai, G.; Buldt, G. *J. Mol. Biol.* **1990**, 214, 15-19.
35. Hildebrandt, P.; Stockburger, M. *Biochemistry* **1984**, 23, 5539-5548.
36. Harbison, G.S.; Roberts, J.E.; Herzfeld, J.; Griffin, R.G. *J. Am. Chem. Soc* **1988**, 110, 7221-7227.
37. De Groot, H.J.M.; Smith, S.O.; Courtin, J.; van der Berg, E.; Winkel, C.; Lugtenburg, J.; Griffin, R.G.; Herzfeld, J. *Biochemistry* **1990**, 29, 6873-6882.
38. Deng, H.; Huang, L.; Callender, R.; Ebrey, T. *Biophys. J.* **1994**, 66, 1129-1136.
39. Fischer, W.B.; Sonar, S.; Marti, T.; Khorana, H.G.; Rothschild, K.J. *Biochemistry.* **1994**, 33, 12757-12762.
40. Dencher, N.A.; Sass, H.J.; Buldt, G. *Biochim. Biophys. Acta* **2000**, 1460, 192.
41. Kandori, H. *Biochim. Biophys. Acta* **2000**, 1460, 177.
42. Nina, M.; Smith, J.C.; Roux, B. *J. Mol. Struc. (THEOCHEM)* **1993**, 286, 231-245.
43. Nina, M.; Roux, B.; Smith, J.C. *Biophys. J* **1995**, 68, 25-39.
44. Otto, H.; Marti, T.; Holz, M.; Mogi, T.; Lindau, M.; Khorana, H.G.; Heyn, M.P. *Proc. Natl. Acad. Sci. U.S.A.* **1989**, 86, 9228-9232.
45. Holz, M.; Drachev, L.A.; Mogi, T.; Otto, H.; Kaulen, A.D.; Heyn, M.P.; Skulachev, V.P.; Khorana, H.G. *Proc. Natl. Acad. Sci. U.S.A* **1989**, 86, 2167-2171.
46. Butt, H.J.; Fendler, K.; Bamberg, E.; Tittor, J.; Oesterhelt, D. *EMBO J.* **1989**, 8, 1657-1663.
47. Gerwert, K.; Hess, B.; Soppa, J.; Oesterhelt, D. *Proc. Nat. Acad. Sci., U.S.A* **1989**, 86, 4943-4947.

48. Spassov, V.Z.; Luecke, H.; Gerwert, K.; Bashford D. *J Mol Biol.* **2001**; 312, 203.
49. Baudry, J.; Crouzy, S.; Roux, B.; Smith, J.C. *J. Chem. Inf. Comp. Sci.* **1998**, 37, 1018-1024.
50. Baudry, J.; Crouzy, S.; Roux, B.; Smith, J.C. *Biophys. J.* **1999**, 76 1909-1917.
51. Crouzy, S.; Baudry, J.; Smith, J.C.; Roux, B. *J. Comput. Chem.* **1999,** 20, 1644.
52. Simon, S.; Aalouach, M.; Smith J.C. *Faraday Discuss.* **1998**, 111, 95-102.
53. Jones, C.R.; Sikakana, C.T.; Hehir, S.; Kuo, M.-C.; Gibbons, W.A. *Biophys. J.* **1978**, 24: 815-832.
54. Hull, E.; Karlsson, R.; Main, P.; Woolfson, M.M.; Dodson, E.J. *Nature* **1978**, 75, 206-207.
55. Mihailescu, D.; Smith, J. C. *J. Phys. Chem. B* **1999**, 9, 1586-1594.
56. Datema, K.P.; Pauls, K.P.; Bloom, M. *Biochemistry* **1986**, 25, 3796-3803.
57. Zidovetzki, R.; Banerjee, U.; Harrigton, D.W.; Chan, S.I. *Biochemistry* **1988**, 27, 5686-5692.
58. Prenner, E.J.; Lewis, R.N.A.H.; Neuman, K.C.; Gruner, S.M.; Kondejewski, L.H.; Hodges, R.S.; McElhaney, R.N. *Biochemistry* **1997**, 37, 7906-7916.
59. Higashijima, T.; Miyazawa, T.; Kawai, M.; Nagai, U. *Biopolymers* **1986**, 25, 2295-2307.
60. Roux B. *Acc. Chem. Res.* **2002**, 35, 366.
61. Katsu, T.; Kobayahi, H.; Hirota, T.; Fujita, Y.; Sato, K.; Nagai, U. *Biochim. Biophys. Acta* **1987**, 899, 159-170.
62. Portlock, S.H.; Clague, M.J.; Cherry, R.J. *Biochim. Biophys. Acta* **1990**, 1030, 1-10.

Indexes

Author Index

Alexov, E. G., 93
Balabin, Ilya A., 107
Ceccarelli, Matteo, 37
Cukier, Robert I., 145
Georgievskii, Yuri, 119
Gunner, M. R., 93
Kim, Jongseob, 119
Knapp, Ernst-Walter, 71
Marchi, Massimo, 37
Mihailescu, Dan, 175

Onuchic, José Nelson, 119
Pomès, Régis, 159
Rabenstein, Björn, 71
Smith, Jeremy C., 175
Souaille, Marc, 37
Stuchebrukhov, Alexei A., 119
Ullmann, G. Matthias, 175
Wheeler, Ralph. A., 1, 51
Zerner, Michael C., 7
Zheng, Xuehe, 119

Subject Index

A

Adenosine triphosphate. *See* ATP synthesis
Alternating access model, 163
Antenna molecules, chlorophyll pigments, 8
Antenna system, 72
Atomic charges, dielectric constant, and dielectric boundary, electrostatic energies, 84–87
Atomic partial charges calculation, CHELPG-like method, 78
ATP synthesis, 52

B

B band in porphyrins, extinction coefficients, spectroscopy, 14–16
Bacterial photosynthetic reaction center, electron transfer energetics, 71–92
Bacterial photosystems, 52–54
Bacteriochlorophyll derivative, DFT *ab initio* calculation, 43
Bacteriochlorophyll dimer. *See* Special pair entries
Bacteriochlorophylls, parament refinement in modeling, 43–45
Bacteriorhodopsin
 hydration and conformational thermodynamics, 178–180
 light-driven proton pump, 178–180
 transmembrane ion pump, 4
p-Benzoquinone
 energy changes in one-electron reduction, 58–62
 reduction in photosynthetic reaction center, thermochemistry, 51–69
p-Benzoquinones and semiquinone anions, properties, 55–58
p-Benzoquinones in photosynthesis, structures, 53f
Binding site, dark-adapted and light-exposed crystal structures, *Rhodobacter sphaeroides*, 73f
Binding sites, electron transfer process
 early calculations, 73–76
 protonation patterns, calculations, 76–89
Bioenergetics, molecular-level process, 1–2f

C

Charge distribution in reaction center, self-consistent reaction field model, 27–30
CHELPG-like method, calculation of atomic partial charges, 78
Chlorophyll a, absorption spectrum 18f
Chlorophyll molecular structure, comparison to porphyrins, 8–9f
Chloroplasts, role in photosynthesis, 52
Chromophore modeling, 42–45
Chromophores
 bacterial photosynthetic reaction center, 37
 density functional theory calculations, 42–45
 photoexcitation modeling by spin-boson model, 39–40

Closed-shell molecule, self-consistent field configuration, calculation, 11, 13f–14
Clusters near Q_B, behavior, 99–100
Coherence parameter, 109–110, 113
Conduction and leakage mechanisms, proton relay by water wires, 168–170
Conformation effects in electron transfer, 99–100
Coordinates and molecular composition, discrepancies, electrostatic energies, 83–84
Coupled chromophore model, 19, 22f
Current density calculations, Hartree-Fock calculations, 122–124
Cyanobacteria photosystems, 52

D

Dark adaptation, bacteriorhodopsin, 180
Density functional theory
  calculations, charges high-spin non-heme iron, 78
  chromophore calculations, 42–45
  methyl bacteriopheophorbide, technique, 43
Detergent micelle, simulation of a reaction center, 46–49
DFT. *See* Density functional theory
Dielectric boundary and constant, atomic charges, electrostatic energies, 84–87
Dipole moments, polarizable matter around chromophores, 26–28
Discrepancies in electrostatic energies, reasons, 81–89
Dissociative proton-coupled electron transfer, 151–154
Donor-Bridge-Acceptor systems, calculations, long-distance electron tunneling, 119–144
Dynamic amplification, 113

Dynamics, electron transfer pathways in redox proteins, 107–117

E

Effective tunneling coupling calculations, 109–110
Electron affinity calculations for *p*-benzoquinones, B3LYP HF/DF method, 59–62
Electron, proton, and energy transfer in molecular bioenergetics, 1–6
Electron reduction potentials for *p*-benzoquinones in water, 62t
Electron transfer
  analysis in proteins, protein pruning, 126, 128f–129
  energetics in bacterial photosynthetic reaction center, 71–92
  Q_A^- to Q_B in reaction center proteins, modeling, 93–105
  proton-coupled reactions, 145–158
  ruthenium-modified azurin system, 125–129
  See also Reaction centers
Electron transfer pathways
  dominant pathways, 111–115
  redox proteins, dynamics, 107–117
Electron tunneling, long-distance, *ab initio* calculations, 119–144
Electronic Hamiltonian, use in electron transfer rate calculations, 110
Electrostatic energies, reasons for discrepancies, 81–89
Energy-transducing membranes, proton transport, 159–160
Energy-transducing proteins and peptide membrane interactions, computer models, 175–186
Et. *See* Electron transfer
Ewald re-summation technique, 45
Exergonic proton transfer following electron transfer, 154–157

F

Fermi Golden rule, reaction rate in the weak coupling limit, 109–110
Franck-Condon factors
  biological electron transfer reaction rates, 107, 109, 115
  initial (final) state proton vibronic wavefunctions, 149–150
  OTO Approximation, 123, 124
Free energy, electron transfer, Q_A^- to Q_B, 79–80, 96–103

G

Gaussian program with LanL2DZ basis set, tunneling flow calculations, 131
Gramicidin A, proton-conducting channels, 159, 167–168
Gramicidin S, peptide–membrane molecular dynamic studies, 181–183
Green's functions, molecular dynamics computation, 110, 115
Grotthuss mechanism, 160–161, 166, 167, 169

H

Halobacterium halobium, bacteriorhodopsin in membrane, 178
Hartree-Fock calculations, current density calculations, 122–124
Hartree-Fock density functional calculations, p-benzoquinones, 51–69
Heme a, model system, nature of tunneling orbitals, 138–141f
Heme molecular structure, comparison to chlorophyll, 8–9f
Highest occupied molecular orbital. *See* HOMO
HOMO, comparison to tunneling orbital, 138–140
Hund's rule, 14
Hydration and conformational thermodynamics, bacteriorhodopsin, 178–180
Hydration free energy differences, p-benzoquinones in water, 61–62

I

INDO/s model, semi-empirical scheme in calculations on special pair, 16, 19–26
Initial photochemical event in photosynthesis, quantum chemical view, 7–35

K

Karlsberg program
  protonation pattern calculations, proteins, 78
  quinones, redox potential calculations, 78–80
Kohn-Sham orbital expansion, chromophores, 43

L

Laurel dimethylamine oxide
  DFT calculations, 43–44
  reaction center simulation, *Rhodobacter sphaeroides*, 46–49
Leakage and conduction mechanisms, proton relay by water wires, 168–170
Lennard-Jones parameters, refinement, 43–45
Light absorption and initial transfer in photosynthesis, 8–9f
Light-driven proton pump, bacteriorhodopsin, 178–180

Lipid bilayers, interactions with gramicidin S, 181–183

Lippincott-Schroeder gas-phase surface, phenol amine cation radical, 152f–153

Long-distance electron tunneling in proteins, *ab initio* calculations by tunneling currents method, 119–144

M

Magnesium chlorin, molecular orbital calculations from spectroscopy, 16–17f

Magnesium etioporphyrin, absorption spectrum, 18f

Marcus theory, rate of biological electron transfer reactions, 107–108

MCCE method. *See* Multi Conformation Continuum Electrostatics

Membrane proteins, proton relay, 159–173

Methods for molecular dynamics, 45–46

Methyl bacteriochlorophyll *a* normal modes, 44f

Methyl bacteriopheophorbide, density functional theory technique, 43

Model charge transfer system, tunneling transition, 129–134f

Modeling by spin-boson model, photoexcitation, chromophores, 39–40

Modeling electron transfer from Q_A^- to Q_B in reaction center proteins, 93–105

Modeling energy-transducing proteins and peptide membrane interactions, 175–186

Modeling first electron transfer in reaction centers, 95–96

Modeling in reaction center protein simulation, 37–50

Molecular bioenergetics, electron, proton, and energy transfer, 1–6

Molecular dynamics
 algorithms, 45–46
 computation, 110–111
 simulations showing Q_B movement, 178

Monte Carlo sampling technique, 78

Multi Conformation Continuum Electrostatics method, first electron transfer, 95–96

Multiple timestep molecular dynamics algorithms, 45

Mutagenesis in experimental structure-function work, 176–178

Mutations, effects on free energy determination, 100–104

N

Non-polar water-filled channels, 165–166

Nuclear coupling, P → P* transition, 38–42

O

One-electron reduction, *p*-benzoquinones, energy changes, 58–62

One Tunneling Orbital Approximation. *See* OTO Approximation

OTO Approximation, 123–125

P

P → P* transition, nuclear coupling, 38–42
 See also Special pairs

Pathways method, tunneling mechanism in proteins, 108, 113–115

Pcet. *See* Proton-coupled electron transfer

Peptide-membrane interactions, 180–183

Photoexcitation modeling by spin-boson model, chromophores, 39–40

Photosynthesis
 initial photochemical event, quantum chemical view, 7–35
 primary events in bacterial photosynthetic reaction centers, 72
 summary chemical reaction, 52
 summary stoichiometry, 7

Photosynthetic reaction center
 electron transfer cofactors, 3–4
 purple bacteria, 176–178
 Rhodopseudomonas viridis, 8, 10*f*–12*f*
 thermochemistry, *p*-benzoquinone reduction, 51–69

Photosystem II oxygen-evolving complex, charge transfer chain, 146–147

Plant photosystems, mechanism, 52

Plumbing, hydrogen-bonded networks, Grotthuss mechanism, 160–161

Poisson-Boltzmann equation, in free energy evaluation, 150

Polarizable matter around chromophores, dipole moments, 26–28

Porphyrins, molecular orbital calculations from spectroscopy, 14–16

Protein changes between $Q_A^-Q_B$ and $Q_A Q_B^-$ states, 97–99

Protein pruning, electron transfer analysis in proteins, 126, 128*f*–129

Proteins in bioenergic process and peptide-membrane interactions, models, 175–186

Proton-conducting channels, 164–170

Proton-coupled electron transfer reactions
 concerted reaction mechanism, 148–150
 consecutive reaction pathways, 148
 dissociative final state, 151–154
 general schemes, 146–151
 strongly exergonic proton transfer, 154–157

Proton plumbing and pumping, molecular level, 164

Proton pumping, thermodynamics and kinetics, 161–163

Proton relay
 leakage and conduction mechanisms by water wires, 168–170
 membrane proteins, 159–173

Proton uptake as function of pH in Q_B reduction, 178

Protonation
 electron transfer, Q_A^- to Q_B, 96–97
 pattern calculations, proteins, 78
 residue near Q_B, 80–81

Q

Q band in porphyrins, spectroscopy, 14–16

Quantum chemical view, initial photochemical event in photosynthesis, 7–35

Quantum-mechanical flux, model system tunneling transition, 132*f*–133*f*

Quinones, redox potential calculations, 78–80

R

Reaction center model, improvement, 42–49

Reaction center protein simulation, molecular modeling, 37–50

Reaction center proteins, modeling electron transfer from Q_A^- to Q_B, 93–105

Reaction centers
  bacterial photosynthetic, electron transfer energetics, 71–92
  calculated spectroscopy, *Rhodopseudomonas viridis*, 26–30f
  electron transfer energetics, *Rhodobacter sphaeroides*, 79–80
  functional properties, *Rhodobacter capsulatus*, 94
  photosynthetic, *Rhodopseudomonas viridis*, 8, 10f–12f
  simulation, detergent micelle, 46–49
  See also Electron transfer
Redox potential calculations, quinones, 78–80
Redox proteins, electron transfer pathways, dynamics, 107–117
Residue protonation near Q_B, 80–81
Residues in cluster, role in electron transfer, 103–104
Retinal equilibrium in dark adaptation, 180
Rhodobacter capsulatus
  functional properties, reaction centers, 94
  site-directed mutagenesis, 176–177
Rhodobacter sphaeroides
  binding site, dark-adapted and light-exposed crystal structures, 73f
  electron transfer energetics, reaction center, 79–80
  molecular modeling, chromophores, reaction center, 37–50
  nuclear coupling in P → P* transition, 38–42
  photosystems, 53–55
  reaction center proteins, modeling electron transfer from Q_A^- to Q_B, 93–105
  reaction center simulation, laurel dimethylamine oxide, 46–49
  ubiquinone reduction, photosynthetic reaction center, 54–55
Rhodopseudomonas viridis
  photosynthetic reaction center, 8, 10f–12f
  photosystems, 53–55
  reaction center calculated spectroscopy, 26–30f
  special pair, model dimer, calculated spectroscopy, 19–26
Ruthenium-modified azurin system, electron transfer, 125–129

S

SCF. *See* Self-consistent field method
Schiff base-water interactions, 179–180
Schrödinger equation, molecular, 11, 157
Self-consistent field method
  configuration calculation, closed-shell molecule, 11, 13f–14
  reaction model, charge distribution in reaction center, 27–30
Solutions to LPB equation, electrostatic energies, discrepancies, 82
Soret band in porphyrins. *See* B band in porphyrins
Special pairs
  charge transfer cycle, bacteria reaction centers, 53–54
  excitation in photosynthesis in purple bacteria, 37–38
  Rhodopseudomonas viridis, calculated spectroscopy, model dimer, 19–26
  See also P → P* transition, nuclear coupling; Ubiquinone-10
Survival probability molecular dynamics scheme, 157

T

Thermochemistry, p-benzoquinone reduction in photosynthetic reaction center, 51–69

Thermodynamic and kinetic control, proton pumping, 162f–163

Thermodynamic stability, water molecules in membrane proteins, 178–180

Transfer matrix element, donor and acceptor diabatic states, calculations, 135–138

Tunneling currents *ab initio* calculations, long-distance electron tunneling, 119–144

Tunneling currents in tunneling transition, method, 120–122

Tunneling mechanism in proteins, *Pathways* predictions, 108

Tunneling orbital, comparison to canonical HOMO, 138–140

Tunneling path in two-dimensional electron-proton tunneling space, 148–149f

Tunneling transition in model charge transfer system, 129–134f

U

Ubiquinone-10
  Q_A^- to Q_B, electron transfer energetics, 71–92
  Q_B reduction, proton uptake as function of pH, 178
  See also Special pairs

Ubiquinone binding in photosynthetic reaction center, 51–69

Ubiquinone reduction, photosynthetic reaction center, 54–55

W

Water molecules in buried cavities in membrane proteins, 178–180

Water-Schiff base interactions, 179–180

Water wires, leakage and conduction mechanisms in proton relay, 168–170

X

X-ray diffraction studies, photosynthesis reaction centers, 54

Z

Zig-zag tunneling path, two-dimensional electron-proton tunneling space, 148–149f

Zundel cations, protonated water dimers, 166